D0874734

Wetlands

EXPLORING ENVIRONMENTAL CHALLENGES
A MULTIDISCIPLINARY APPROACH

SERIES EDITORS
Sharon L. Spray, Assistant Professor of Political Science Elon University
Karen L. McGlothlin, Assistant Professor of Biology The University of the South

ABOUT THE SERIES
Exploring Environmental Challenges: A Multidisciplinary Approach is a series of short readers designed for introductory-level, interdisciplinary environmental sciences, or environmental studies courses. Each reader, focused on a single, complex topic of environmental concern, outlines the concepts, methods, and current research approaches used in the study of that particular environmental challenge from six distinct fields of study in the natural sciences, social sciences, and humanities. This approach enables students and faculty alike to become familiar with a topic from perspectives outside their own training and to develop a broader appreciation of the breadth of efforts involved in investigating select, complex environmental issues.

TITLES IN SERIES
Global Climate Change, Sharon L. Spray and Karen L. McGlothlin
Loss of Biodiversity, Sharon L. Spray and Karen L. McGlothlin
Wetlands, Sharon L. Spray and Karen L. McGlothlin

Exploring Environmental Challenges

A MULTIDISCIPLINARY APPROACH

Wetlands

EDITED BY
Sharon L. Spray & Karen L. McGlothlin

ROWMAN & LITTLEFIELD PUBLISHERS, INC.
Lanham • Boulder • New York • Toronto • Oxford

ROWMAN & LITTLEFIELD PUBLISHERS, INC.

Published in the United States of America
by Rowman & Littlefield Publishers, Inc.
A wholly owned subsidiary of The Rowman & Littlefield Publishing Group, Inc.
4501 Forbes Boulevard, Suite 200, Lanham, Maryland 20706
www.rowmanlittlefield.com

PO Box 317; Oxford; OX2 9RU, UK

British Library Cataloguing in Publication Information Available

Library of Congress Cataloging-in-Publication Data

Wetlands / edited by Sharon L. Spray and Karen L. McGlothlin.
        p. cm. — (Exploring environmental challenges)
Includes bibliographical references and index.
   ISBN 0-7425-2568-6 (alk. paper) — ISBN 0-7425-2569-4 (pbk. : alk.
paper)
   1. Wetland ecology. 2. Wetland management. 3. Wetlands. I.
Spray, Sharon L. II. McGlothlin, Karen L. (Karen Leah) III. Series.
   QH541.5.M3W4655 2004
   333.91'816—dc22

                                                    2003020408

Printed in the United States of America

∞™ The paper used in this publication meets the minimum requirements of American
National Standard for Information Sciences—Permanence of Paper for Printed Library
Materials, ANSI/NISO Z39.48-1992.

Contents

Contributors

Dr. John Callaway received his Ph.D. from Louisiana State University in 1994. He is an assistant professor of environmental science at the University of San Francisco. His research expertise is in wetland restoration, specifically wetland plant ecology and sediment dynamics. He teaches courses in applied ecology, restoration ecology, and environmental sciences. He serves on a number of national and international advisory panels on wetland and coastal issues and is currently the editor of *Madroño*, the journal of the California Botanical Society.

Dr. Stephen Faulkner received his M.S. in forestry from Louisiana State University and his Ph.D. in wetland biogeochemistry from Duke University. He is a wetland research ecologist with the U.S. Geological Survey National Wetlands Research Center in Lafayette, Louisiana, where his primary research interests are in climate change, carbon and nutrient biogeochemistry, and ecosystem restoration. He also serves as an associate editor for the journals *Wetlands* and *Urban Ecology*.

Dr. Mary Hague received her M.A. and Ph.D. in political science from Boston College and was most recently an associate professor of politics at Juniata College. Her research interests include environmental policy and policy history. She has presented several papers on environmental policy, especially addressing the Clear Water Act. Her latest publications include a chapter examining the creation of municipal parks in the late 1800s. She is currently working on a project examining sanitation and environmental reforms of the Progressive era.

Dr. Karen McGlothlin received her M.S. in biological sciences from East Tennessee State University and her Ph.D. in zoology from Clemson

University. She is an assistant professor of biology at the University of the South, in Sewanee, Tennessee, and teaches courses in invertebrate zoology, island ecology, entomology, and developmental biology. She is also an active participant in the interdisciplinary Environmental Studies concentration and in the Island Ecology Program.

Dr. William B. Meyer received a B.A. in geology and history from Williams College and a Ph.D. in geography from Clark University. He has held research appointments at Clark and at Harvard University's Kennedy School of Government and is the author of *Human Impact on the Earth* and *Americans and Their Weather: A History*.

Dr. Thomas Michael Power is professor of economics and chairman of the Department of Economics at the University of Montana, where he specializes in natural resource economics and regional economic development. He received his Ph.D. in economics from Princeton University and is the author of six books, including *Post-Cowboy Economics: Pay and Prosperity in the New American West* and *Lost Landscapes and Failed Economies: The Search for a Value of Place*.

Dr. Joel Snodgrass received his M.S. in zoology from the University of Central Florida and his Ph.D. in ecology from the Institute of Ecology at the University of Georgia. He is an assistant professor of conservation biology at Towson University, in Towson, Maryland. His research involves the influence of human land development on aquatic systems and the biology of aquatic organisms. He teaches courses in ecology and evolution, conservation biology, and landscape ecology.

Dr. Sharon L. Spray is an assistant professor of political science and environmental studies at Elon University in Elon, North Carolina. She earned her Ph.D. from the Claremont Graduate School in Claremont, California. She teaches courses in American politics and international and domestic environmental policy. She also works as an associate of Elon University's Center for Environmental Studies.

Preface

Sharon L. Spray
and
Karen L. McGlothlin

Exploring Environmental Challenges: A Multidisciplinary Approach is a series of short readers designed for introductory-level, interdisciplinary environmental sciences or environmental studies courses. Each reader, focused on a single, complex topic of environmental concern, outlines the concepts, methods, and current research approaches used in the study of that particular environmental challenge from six distinct fields of study in the natural sciences, the social sciences, and the humanities. This approach enables students and faculty alike to become familiar with a topic from perspectives outside their own training and to develop a broader appreciation of the breadth of efforts involved in investigating select, complex environmental issues. This series was developed to facilitate interdisciplinary teaching in environmental studies programs by acknowledging that different disciplines bring distinctly different perspectives to the table and that scholars trained in those fields are best suited to explain these perspectives. The texts in this series are designed to assist faculty trained in a traditional social science, natural science, or humanities field to venture into areas of research outside their own training.

In the past decade, a rapidly increasing number of institutions of higher education across the country have developed a wide variety of interdisciplinary programs in both environmental science and environmental studies. While many of these programs are centered primarily within the science curriculum, more and more institutions are strengthening their environmental sciences and environmental studies majors, minors, and concentrations by adding courses from both the social sciences and the humanities. The importance of integrating information from a variety of disciplines, including the sciences, the social sciences, and the humanities, has been recognized and considered in the design and revision of environmental curricula. Liberal arts institutions, in particular, are moving toward the development of inter- or multidisciplinary approaches as a basis for their environmental programs. These approaches are as varied as the institutions themselves. While many programs offer team-taught courses to provide true interdisciplinary approaches, others are built around a series of courses from across curricula that address environmental topics. The foundation for and value of such programs is the recognition that complex environmental challenges will necessarily require strengthening the interface between the social sciences, humanities, and natural sciences if we hope to find productive ways of addressing these issues.

The concept for this series grew from discussions that emerged during the planning and development stages of an environmental studies program at the University of the South during the late 1990s. One of the points agreed on during our discussions was that all students enrolled in the environmental studies programs would be required to take an introductory course that would be interdisciplinary in nature and taught by a team of professors from the natural and the social sciences. In our case, that meant a teaching collaboration between a zoologist and a political scientist.

During the course of our conversations and explorations into the available literature, we found ourselves feeling a bit overwhelmed at the thought of teaching a truly interdisciplinary course. We felt that it would be difficult, at best, to hold classroom discussions concerning different concepts and approaches to the studies of various environmental topics from a variety of academic disciplines with our training centered in our particular fields. After much discussion, we decided that a series of edited readers, with each volume focusing on a single, complex environmental topic and chapters written by experts in various fields, would be of great use to students and faculty involved in interdisciplinary environmental studies programs. Thus, we began the development of this multidisciplinary series of readers on environmental challenges.

During the conceptualization stage of this project and later, during the proposal review stage, the issue of "multidisciplinary" versus "interdisciplinary" teaching surfaced repeatedly. These two terms are frequently heard in discussions pertaining to environmental studies programs and often are used interchangeably. For this reason, we feel that it is important that these two terms be defined, providing, it is hoped, clarification for some and reassurance to others that a volume written from a multidisciplinary perspective can be used in an interdisciplinary course.

When we speak of multidisciplinary perspectives, we are referring to distinct disciplinary approaches to the study of a particular topic. Such perspectives do not preclude the integration of knowledge or material from other fields, but the interpretation of the information reflects a particular disciplinary perspective. We view this as a matter of disciplinary depth. As scholars, we necessarily cross the boundaries of knowledge and scholarship from other fields, but most of us have more depth in the field in which we received our academic training. Consequently, we interpret information through particular theoretical perspectives tied to our disciplinary training.

We view interdisciplinary teaching as the attempt at balanced integration of material from multiple disciplines. This, however, is a difficult goal when studying environmental issues. Most texts written about specific environmental issues reflect heavy bias toward the natural sciences with some discussion of policy and economics, or, alternatively, the focus may be in the opposite direction with an emphasis on policy and economics and limited discussion of science. More problematic is that many of the available texts fail to incorporate in any meaningful way the work of humanists, anthropologists, or sociologists—areas we believe are essential for understanding complex environmental challenges.

The texts are purposefully balanced with half the chapter contributions from the natural sciences and an equal number of chapters contributed from scholars in the humanities or social sciences. Each chapter identifies important concepts and theoretical perspectives from each field, and each includes a supplemental reading list to facilitate additional study. We envision these texts to be the foundation for introductory environmental studies courses that examine environmental topics from multiple perspectives or other courses that seek an interdisciplinary focus for the study of environmental problems. Because we anticipate that students from a variety of majors, both science and nonscience, will use these texts, the chapters are designed to be understandable to those with little familiarity of the topic or the field about which it was written.

The series is not neutral in its basic premise. The various topics in the series were chosen because we believe that the topics addressed are environmental challenges that we want students to better understand and we hope work toward future solutions. Individual authors, however, were asked to provide objective presentations of information so that students and faculty members could form their own opinions on how these challenges should be addressed. We care deeply about the environment, and we hope that this series serves to stimulate students to take the earth's stewardship seriously and promote a better understanding of the complexity of some of the environmental challenges facing us in this new century.

Introduction

Sharon L. Spray
and
Karen L. McGlothlin

What comes to mind at the mention of the term "wetlands"? For many people, it would be an image of a typical coastal marsh system: an area teeming with a variety of animal life and characterized by soft, sticky mud and emergent vegetation such as rush (*Juncus*) and cordgrass (*Spartina*). In reality, the term "wetlands" encompasses not only coastal marshes but highly diverse inland freshwater ecosystems as well. In fact, of the estimated 105.5 million acres of wetlands remaining in the contiguous United States in 1997, 95% were inland freshwater wetlands (Dahl 2000). Because of their diversity, many regions may not be easily recognizable as a "wetland" outside the scientific community.

The term "wetlands" is actually a broad term that is generally used by scientists to refer to areas that are distinguished by the presence of standing water during all or part of the year, leading to the formation of **hydric soils** and the presence of **flora** and **fauna** that are adapted to surviving under the unique, saturated conditions found in wetlands (Faulkner, this volume; Mitsch and Gosselink 2000). Surprisingly, there is currently no generally accepted classification scheme for wetlands to guide research or inventories of these valuable ecosystems. Some classification schemes have been determined on the basis of hydrologic flow regimes, some based on vegetation types, and some on **hydrogeomorphology**. Classification schemes also have been developed for wetlands in different regions of the country and for different states (for an example of additional classification schemes, see www.h2osparc.wq.ncsu.edu/info/wetlands/types3.html). The U.S. Environmental Protection Agency (EPA) classifies wetlands on the basis of four major types—marshes, swamps, bogs, and fens—providing a basic but simplified scheme for understanding the diversity of wetland ecosystems (www.epa.gov/owow/wetlands/types).

Marshes are classified as wetlands because the soil in these regions are either periodically or continually inundated with water and they have a characteristic flora composed of emergent vegetation species adapted to life in saturated soils. **Tidal marshes** are located in areas that are alternately flooded and dried as a result of ocean tides but are protected from heavy wave action along coasts worldwide. These types of wetlands may be freshwater, **brackish**, or **saline**. Unlike tidal marshes, **nontidal marshes** are typically inland freshwater or brackish marshes that may periodically dry up. Using the EPA classification scheme, nontidal marshes include vernal pools (shallow depressions found in grasslands or forests that are typically flooded only in the winter or early spring), wet meadows (grasslands that are characterized by soils that become waterlogged after precipitation events but retain no standing water during most of the year), prairie potholes (small, shallow ponds that form in depressions caused by glaciation in the Dakotas, Iowa, and the prairies of the central Canadian provinces), and playa lakes (shallow ponds found in the arid to semiarid regions of the southwestern United States, similar to prairie potholes but of different geologic origin).

Swamps are wetlands that are dominated by woody plants, either shrubs or trees, and characterized by saturated soils during the growing season and standing water at other times of the year. The Great Dismal Swamp, located in northeastern North Carolina and in southeastern Virginia, is a classic example. This 109,000-acre forested wetland includes a swamp forest characterized by bald cypress and dominated by tupelo, red maple, and black gum and provides habitat for a wide variety of animal life, including white-tailed deer, river otter, bats, raccoon, mink, black bear, bobcat, snakes, turtles, and over 200 species of birds. **Bogs** are a distinctive category of wetlands often quite different in appearance from marshes and swamps and characterized by highly different soils and vegetation. Bogs are discernable by the presence of peat deposits, acidic water, and a typical ground cover of sphagnum moss. Bogs are known as **precipitation-dominated wetlands** because they receive most of their water inputs from precipitation as opposed to groundwater, streams, or runoff. Specific types of bogs include **pocosins**, which are evergreen-shrub bogs located in the southeastern United States, and **northern bogs**, found in the Northeast and Great Lakes regions of the United States. The final category of wetlands designated by the EPA includes the **fens**, which are also distinguishable for their peat accumulations. Unlike bogs, however, fens receive their water supplies from groundwater inputs, as opposed to precipitation (**groundwater-dominated wetlands**). Fens also differ from bogs in that they are characterized by less acidic water and higher nutrient levels.

No matter the classification scheme used, there is no disputing that the functions of wetlands in our environment are varied and exceedingly important. Scientists contributing to this book highlight in their chapters the value of the services provided by wetlands, noting how these complex ecosystems contribute to the earth's overall environmental quality. Wetlands have been referred to as "the kidneys of the earth" and function to filter pollutants through the removal of organic and inorganic nutrients and toxins from the water that moves through them. This function results from several characteristics of wetlands, including a large area in which sediments are in contact with water, resulting in a high level of sediment–water exchange; increased rates of mineral uptake by wetlands vegetation as a result of high primary productivity; and the ensuing burial of those minerals in sediments when the plants die. Wetlands also facilitate a decrease in flow velocity as running waters enter wetlands, resulting in sediments and any chemicals that may be bound to sediment particles to drop out of the water column (for additional details, see Mitsch and Gosselink 2000). Wetlands provide flood control by serving as reservoirs for storm runoff, prevent coastal erosion, and provide critically important habitat for plants and wildlife, including many of the nation's endangered species. Wetlands, which comprise an estimated 3.5% of the land area of the United States, were the primary habitat of about 50% of the 209 species of animals listed as endangered in 1986 (Mitsch and Gosselink 2000). Wetland ecosystems are neither easily regenerated nor artificially created, making their preservation critically important to our nation's environmental health.

In 1989, the first Bush administration set forth a "no net loss" of wetlands policy to preserve wetlands throughout the United States, but there has continued to be a decline at the rate of approximately 57,000 acres annually, with many more additional acres of wetlands degraded to such degree that ecosystem functions are compromised. Wetland degradation and losses may be attributed to a host of pressures, brought about primarily as a result of human activities (www.epa.gov/owow/wetlands/factsheets/threats.pdf). Some of the major threats to wetlands have been the result of changes in wetlands hydrology, which is the water regime required to maintain a functional wetland (see Faulkner, this volume). Any variation in a wetland's hydrology puts tremendous pressures on all the organisms inhabiting that area and, in turn, compromises the functions of that wetland. Hydrologic alterations may arise from sources as varied as the filling in or drainage of wetlands for development and agriculture to the channelization or dredging of streams for navigation. Wetlands degradation and/or destruction may also result from the introduction of excess pollution into the ecosystem. While one of the functions of wetlands is pollution reduction, the ability of a wetland to perform this function may be exceeded, especially in areas where activities such as fertilizer and pesticide runoff, sedimentation, or heavy-metal contamination is prevalent. Vegetation damage may also prove to be detrimental to wetland function. This damage may arise as a result of either of the previously mentioned threats, hydrologic alteration, or excess pollution but may also result from the introduction of invasive species that may outcompete native wetlands plants, grazing by domestic animals, or the removal of vegetation for peat mining (U.S. Environmental Protection Agency 2003). Because wetland degradation and loss occurs as a result of such a diverse list of causal factors, devising effective policy prescriptions remains extraordinarily difficult.

This is neither a science book nor a policy book. This book is intended to introduce students to the study of wetlands with the purpose of providing enough background in their natural ecological functions and policy issues that readers will have a solid understanding of why these ecosystems are necessary to our nation's environmental health and why their governance remains a significant environmental challenge. Contributors to this book discuss wetlands from six disciplinary perspectives: three from the natural sciences and three from the social sciences. Although this book is not a comprehensive overview of all perspectives or important information on wetlands, it does provide background material for a more comprehensive investigation of the science behind and the management of these ecosystems. The disciplines included in this volume are not a reflection of a belief that some fields of study are more important than others. Rather, our goal is to provide readers with a basic tool kit for understanding wetlands ecology and wetlands policy. The chapters are not ordered by importance, and most chapters interrelate in their discussions. We begin with chapters from scientists who provide an overview of many of the key ecological concepts pertinent to understanding wetlands ecology and progress to chapters by social scientists who focus on historical management of wetlands in the United States and important political and economic concepts necessary to understand present and future policy development.

The first chapter of this book, written by biologist Joel Snodgrass, focuses on the current research of wildlife and fisheries biologists. Particular attention is given to freshwater wetlands. The chapter is rich in discussion of ecological concepts necessary for understanding the uniqueness of wetland ecosystems as well as important wetlands fish and wildlife management concepts. This chapter is highly recommended for any reader interested in why the preservation of wetlands in the United States is extremely critical for maintaining our nation's biodiversity.

The specifics of wetland soil chemistry and additional information on wetland hydrology are discussed in chapter 2. In this chapter, biogeochemist Stephen Faulkner breaks down the important chemical processes that occur in wetland ecosystems. This chapter is by far the most technical of all the chapters in this book, but we encourage readers to carefully read this chapter. The chemistry of wetland soils and the functions that wetland environments provide in cleansing groundwater, pollution mitigation, and the maintenance of diverse plant and animal populations is highly important in understanding the value of wetland ecosystems. In policy circles, wetland soil chemistry is perhaps the least understood aspect of wetland ecology but arguably the most important. This chapter is important for understanding discussions of wetlands degradation, wetlands restoration, and wetlands creation.

In the past two decades, policymakers have shifted to wetland creation and wetlands restoration projects to address continuing levels of wetland losses across the nation. Restoration ecologist John Callaway introduces the reader to key concepts in restoration ecology and what scientists have learned about our current ability to create and restore degraded wetlands to natural or pristine conditions. Our nation has dredged, dammed, and drained wetlands for more than two centuries. Whether we can recover from this destruction will depend on our future conservation efforts and our ability to create new and restore previously degraded wetland regions.

The history of our wetland resources is discussed by environmental historian William Meyer in chapter 4. For reasons associated with human health, mechanization of agriculture, transportation, and even stupidity, much of our nation's wetlands were purposefully destroyed in the previous century with the blessing and assistance of the state. To some extent, our past wetlands destruction was about social change, but it was also about a lack of scientific knowledge of wetland functions. Meyer provides an interesting, comprehensive summary of wetlands policy in the United States and points out that even though we may be far more aware of the services that wetlands provide, we are, indeed, more capable now than in the past to quickly convert them to other land uses. Whether we will be able to preserve what is left of our nation's wetlands remains in question.

Contemporary wetlands policy is discussed in chapter 5 by political scientist Mary Hague. Hague walks the reader through the complicated issues associated with wetlands preservation, including the rights of land owners, regulatory federalism, and the policy process. Hague also discusses the implementation and weaknesses of various statutes and the role of the nation's courts in mediating conflicts. This is an especially valuable chapter for anyone who wants to better understand the politics of wetlands preservation.

Chapter 6 discusses the economics of wetlands valuation. Natural and social scientists alike recognize that wetlands conversion projects have increased flooding in some areas of the country, contributed to dramatic declines in fish stocks and waterfowl, and resulted in much higher levels of pollution in drinking water in many areas of the country. Despite our many policies, wetlands regions continue to decline each year. Economists refer to wetlands as public goods—providing services to society that are not divisible into individual units but shared by all. Economist Thomas Power discusses the economic valuation of wetlands and the difficulties we encounter when making decisions about land use in general and wetlands in particular.

The chapters in this book are only introductions to six disciplinary perspectives that help us understand the uniqueness and value of these diminishing regions in our nation. In addition to each chapter, authors have provided additional reading lists so that readers can further explore the issues and concepts raised in this volume. We hope that reading this volume will inspire readers to work more vigorously to protect these regions.

REFERENCES

Dahl, T. R. 2000. *Status and trends of wetlands in the conterminous United States 1986 to 1997.* U.S. Department of the Interior and U.S. Fish and Wildlife Service, Washington, D.C.

Mitsch, William J., and James G. Gosselink. 2000. *Wetlands.* 3rd ed. New York: Wiley.

U.S. Environmental Protection Agency. 2003. www.epa.gov/owow/wetlands.

Wetlands

1

Wetlands Wildlife and Fisheries

A Biological
Perspective of
Wetland
Ecosystems

Joel W. Snodgrass

Wildlife and fisheries biologists are concerned with maintaining and enhancing mainly vertebrate populations, most specifically fish, amphibians, reptiles, birds, and mammals. Both disciplines have historically focused on species of recreational and commercial interest with the goal of maintaining population sizes at levels that will sustain recreational or commercial harvest or both. Traditionally, management was conducted on a species-by-species basis. More recently, wildlife and fisheries biologists have begun to focus on **nongame species** and expand their research and management beyond the single-species approach (Edwards 1989; Mather et al. 1995; Meffe and Carroll 1997), although funding at the state level remains low for these activities (Ross and Loomis 1999). These changes were precipitated by changing public views of wildlife; new regulations and their resulting changes in directives of many local, state, and federal agencies; and the realization of the impacts of habitat destruction on many of the world's **ecosystems** and the commodities and the **ecosystem services** they provide humans. While fisheries and wildlife biologists focus much of their attention on vertebrates, an understanding of plant and invertebrate biology is essential because these organisms provide habitat structure and food for vertebrates. Additionally, some invertebrates found in wetlands and other aquatic systems are of recreational or commercial value. For example, crayfish harvested from wetlands in the southeastern United States are important commodities in these areas (Avery and Lorio 1999).

The value of wetlands for fish and wildlife species has long been recognized. Many game and nongame species utilize wetland habitats during all or a significant portion of their lives, making wetlands critical resources for a multitude of fish and wildlife species. Many birds, amphibians and reptiles, and some mammals forage and reproduce in wetland habitats. Many fish, some amphibians and reptiles, and a few mammals (such as beavers, otters, and muskrats) spend their entire lives in wetlands. Included among these organisms are species of economic value and recreational concern (such as **waterfowl**) and species listed by the federal government as "endangered" or "threatened" (such as wood storks; Boylan and MacLean 1997). With the extensive loss of wetland acreage in the United States (Dahl 1990) and throughout the world, it is not surprising that fisheries and wildlife biologists are concerned with wetland biology, management, and conservation. Furthermore, wetland areas are being created in great numbers in many developing areas as part of storm water management programs; the effects of these systems on fish and wildlife remain poorly understood, and the potential for negative effects exists (Helfield and Diamond 1997).

This chapter focuses on summarizing existing paradigms and major current research efforts in wildlife and fisheries biology with a particular emphasis on wetlands. Wildlife and fisheries biologists draw from ecological concepts in their studies of the basic ecology of species and ecosystems and in crafting and implementing their **management plans**. Therefore, a brief introduction to those ecological concepts most pertinent to wetlands is included. Concepts are illustrated primarily with examples from freshwater wetlands. However, it should be kept in mind that **estuarine** and **tidal wetlands**, such as mangrove swamps, sea-grass beds, and *Spartina* marshes, are important resources for fish and wildlife and are being destroyed at rates equal to or exceeding those of freshwater wetlands (Gosselink and Maltby 1990).

Methods in Wildlife and Fisheries Biology

In general, fisheries and wildlife biologists strive to understand the relationships between organisms and their environment and apply that information in managing fish and wildlife species and the habitats they depend on. At the level of a **population**, fish and wildlife biologists ask, What resources are needed for individuals to survive, grow, and reproduce successfully? For example, what does a specific species eat, where does it nest, what habitats are needed for acquiring food and protection from predators, and what population size is needed to be reasonably confident that it will persist for some extended period of time? Presumably, if these conditions are met, the population will persist, and if they are enhanced, the population will grow in size. These same general ideas can be extended to higher levels of biological organization, such as **communities** and ecosystems. As mentioned in the introduction, the focus of fisheries and wildlife biologists is currently being extended to include higher levels of biological organization. So, fisheries and wildlife biologists may also ask, What **natural disturbance regime** or **patch size** of an ecosystem must be maintained for the ecosystem to continue to function naturally?

Descriptive studies form the basis of knowledge of species, communities, and ecosystems and lead to hypotheses that can be tested using experimental methods. Documenting the natural history of organisms still plays a large role in fisheries and wildlife management. Thus, describing the diet and activities of individual organisms in the field occupies much of the time of fisheries and wildlife biologists. While computers have led to the development of statistically sophisticated techniques of identifying **preferred habitat,** the goals of these studies consist of developing a clear picture of what resources a species needs to complete its life cycle (that is, to survive and reproduce). Studies directed at determining resource needs of animals often involve locating animals by sight, active capture, or passive trapping. Captured or trapped animals may be marked or fitted with radio transmitters so they can be identified or located at a later date.

From an operational standpoint, fisheries and wildlife biologists often identify the resources needed by a species through association of the species with a **habitat type.** Habitat types are identified by the biologist and may include such categories as pine forest, riparian forest, or marsh. Alternatively, the biologist may measure the structure of the habitat in more detail. For example, the soil moisture and types of plant species present may be measured in areas that are used by an animal. Use of specific habitat types or environmental conditions are then compared with the available habitat types or conditions to determine habitat preference. For example, if an animal was found in marsh habitat 75% of the time but marsh habitat made up only 15% of the habitat available in the area being studied, then the biologist might conclude that the animal prefers marsh habitat.

Because the approach to identifying important resources for the animals described here assumes that organisms have the highest rates of reproduction in preferred habitats and because fisheries and wildlife biologists are concerned with the effects of harvest on populations, they are also concerned with the **demographic rates** of populations. Demographic rates include the number of births and deaths in a population over a given time period, often expressed as a rate per female, and the number of **immigrants** and **emigrants** over a given period of time. Demographic rates may be estimated indirectly using the **fecundity** and **age structure** of the population or directly by tracking the activity of individuals within populations. With this information, the preferred habitat of an animal can then be defined as those areas where birthrates are equal to or exceed death rates and emigration. From an applied standpoint, harvest rates resulting in death rates that exceed birthrates will lead to population declines (possibly local extirpation) and, in the worst-case scenario, extinction of the species. Therefore, fisheries and wildlife biologists often devote large amounts of time to monitoring and regulating harvest rates.

Descriptive studies are also important in understanding community and ecosystem structure and dynamics. Most ecologists now view communities and ecosystems as dynamic in nature, responding to external as well as internal forces. Community properties that are often measured include **species richness** and **relative abundance, diversity,** and colonization and extirpation rates. Ecosystem properties include rates of **primary** and **secondary production, nutrient cycling,** flows of water and energy through ecosystems, and disturbance regimes. The methods for measuring properties of communities and ecosystems are often specific to individual taxonomic groups and processes, and their description is beyond the scope of this chapter. General introductions to measuring community properties can be found in introductory ecology texts, such as Ricklefs and Miller (1999) and Smith (1996). Mitsch and Gosselink (2000) provide methods for measuring ecosystem properties in many types of wetland systems.

While descriptive studies provide valuable insight into the ecology of species, communities, and ecosystems, determining cause-and-effect relationships with such data is often problematic. For example, a biologist may observe a negative correlation between water levels in a wetland and the growth of a species of interest. Several other factors may also vary with water level; pH and the availability of food resources may decrease as water levels fall. Are changes in pH or food resource (or some other factors associated with water-level fluctuations) responsible for decreased growth? An experimental approach is needed to answer this question. The experimental approach involves controlling some of the factors while systematically varying one or more other factors of interest. For example, organisms may be grown in laboratory microcosms in which resource abundance and water levels may be held constant while pH is varied over a range that includes pH values observed in the field. If a correlation between pH and growth consistent with that observed in the field is found in the laboratory, it suggests that decreases in pH have a negative effect on growth.

Fisheries and wildlife biologists employ three types of experiments (Diamond 1986): laboratory, field, and natural experiments. Laboratory experiments, as the name suggests, are conducted in the laboratory where conditions can be precisely controlled but are greatly simplified in relation to field conditions. Field experiments involve creating replicate systems or manipulating natural systems in the field and sacrifice precise control of experimental conditions among replicates for the more ecologically complicated and realistic conditions of the field. Natural experiments take advantage of unplanned natural or human perturbations of systems and allow for the study of phenomena that cannot be addressed with laboratory and field experiments. Such factors might include the effects of hurricanes and tornadoes or the conversion of a watershed from forested to residential land use. Because perturbations are not planned in natural experiments, there is little control of variation in conditions among study sites. One type of experiment commonly used in investigations of the ecology of wetland systems is the use of field **mesocosms**. Mesocosms are constructed in the field as small replicates of natural systems. They represent an intermediate level between laboratory experiments and field manipulations of natural systems because they include more of the complexity and environmental variation found in natural systems while reducing variation among replicates (because they are constructed by the investigator).

While fisheries and wildlife biologists gather basic data on the ecology of organisms and ecosystems, they are also interested in using that information to manage these systems. Traditionally, management techniques have consisted of regulating harvest and manipulating habitats to enhance population growth. With the inclusion of nongame species, habitat manipulation has taken on a more prominent role. Habitat manipulations may range from mechanically removing vegetation to conducting controlled burning of ecosystems. Although management plans are formulated on existing knowledge of population biology and ecosystem function, they are implemented in the complex natural environment and may not always produce the desired results. Therefore, it is important for fisheries and wildlife biologists to monitor the effects of their management activities. Depending on management goals, the focus of monitoring efforts may range from determinations of population sizes to the ecological integrity of an ecosystem. Ecological integrity refers to the normal functioning of ecosystems (Karr 1991). Operationally, an ecosystem with ecological integrity exhibits structure and function that is similar to other pristine ecosystems of the same type.

This section provided a brief introduction to some of the methods used by fisheries and wildlife biologists to study and manage wetlands. Application of these methods has led to a general understanding of wetland ecosystem function, the benefits of wetlands to fisheries and wildlife, and the biology of organisms that inhabit wetlands. However, there is still much to learn. The next section provides a summary of our understanding of the ecology of organisms inhabiting wetlands and how this knowledge is being used to manage wetlands for fish and wildlife.

Current Understanding and Research

Natural history studies of many species have resulted in the recognition of a number of associations between organisms and wetland habitats and resources. For some organisms, resources needed to complete the life cycle may all be found within wetland habitats, but for others, wetland resources may only be needed for completion of a portion of the life cycle. For example, many wading birds may reproduce and forage almost entirely in wetlands habitats. In contrast, many insects and amphibians depend on wetland habitats for the development of their eggs and larvae, whereas adults are dependent on terrestrial habitats. Furthermore, some forms (such as larvae of some species and larvae, juveniles, and adults of others) cannot survive long out of water (later these are referred to as "aquatic forms"). Finally, some species may utilize wetland habitats but not require them for the completion of their life cycle. The following discussion concentrates on species that require wetlands for the completion of their life cycle, often referred to as **wetland-dependent species.**

WETLANDS AND WETLAND TYPES

The first general question that arises when studying wildlife and fisheries in wetlands is, What is a wetland? On the surface, this appears to be an easy question to answer; we all are familiar with an area near our home or work that we consider a wetland. However, in the early 1970s, the passage of regulations protecting wetlands (primarily Section 404 of the Clean Water Act) forced wetland scientists to examine this question in depth, and much debate ensued (mainly between scientists and private interest groups). The result was the development of rigid wetland classification systems and specific methods for identifying wetlands by the federal government (Cowardin et al. 1979; Environmental Laboratory 1987) and ultimately the discipline of **wetland delineation** (Tiner 1999). Currently, wetland scientists recognize wetlands based on a combination of components, including the presence of water and the occurrence of unique soils and plant communities (Mitsch and Gosselink 2000). Saturated soil conditions may result from the groundwater table being at or near the surface or from flooding, where water is present on the surface. Unique soils develop in wetland systems because of slow decomposition rates under low dissolved oxygen levels and the subsequent buildup of organic matter. Low oxygen levels in saturated or flooded soils also limit the occurrence of plants to those species adapted to these conditions, referred to as **hydrophytes.** Hydrophytes have the ability to pump oxygen to their roots, where it detoxifies noxious compounds produced by bacteria under limited oxygen conditions. This detoxification mechanism allows hydrophytes to grow and prosper in conditions that are lethal to other plants.

While **hydrology**, soils, and the presence of hydrophytes provide a general definition of a wetland, wildlife and fisheries biologists often specialize in the study of specific types of wetlands defined by their location in the landscape in relation to other deepwater habitats of a more permanent nature, such as lakes and streams. The proximity of wetlands to deepwater habitats controls access to wetlands for totally aquatic organisms such as fish and often influences hydrological regimes and the exchange of materials between the two systems. Specific types of wetlands recognized by wildlife and fisheries biologists include floodplains of large rivers, fringing wetlands of lakes, estuaries and oceans, large expansive freshwater marshes and forested wetlands, depression wetlands, and human-created wetlands. Because floodplains and fringing wetlands are located between deepwater habitats of lakes, estuaries, oceans and river channels, and adjoining terrestrial habitats, these wetlands are characterized by large exchanges of material and organisms with their adjoining deepwater habitats. Additionally, floodplains and fringing wetlands moderate fluxes of material and organisms between terrestrial and deepwater habitats. Floodplains and lake margin wetlands are characterized by seasonal fluctuations in water levels, while fringing wetlands of estuaries and oceans are characterized by daily tidal cycles. Large expansive freshwater marshes and forested wetlands are recognized for their regional significance and consist of many wetland vegetative communities with a small amount of deepwater habitat and islands of terrestrial habitat interspersed. Examples of expansive freshwater marshes and forested wetlands include the Everglades of southern Florida and the Okefenokee Swamp of southern Georgia. Isolated freshwater wetlands are characterized by a lack of seasonal or daily connections with deepwater habitats and include many prairie pothole wetlands, Carolina bays, and mountain bogs. Human-created wetlands are unique because they are often constructed for purposes other than or in addition to providing fish and wildlife habitat. Wetlands are often created to catch storm water runoff from developed areas and remove pollutants from waters before they enter streams, lakes, or estuaries. Because of the rapid runoff of storm waters from impervious surfaces, water-level fluctuations are often greater in magnitude and frequency in created wetlands. Additionally, fish and wildlife utilizing created wetlands may be exposed to the pollutants these wetlands are designed to sequester.

HABITAT STRUCTURE AND FOOD AVAILABILITY AS CRITICAL RESOURCES

As suggested by the focus of fisheries and wildlife biologists on resource needs of organisms outlined earlier, food availability and habitat structure control the types of animals and plants found in any location. The physical and biotic environment of a location determines habitat structure and food availability. The **physical environment** includes moisture levels, temperature, light and nutrient availability, and variation in these factors. The factors most important in shaping the physical environment are the regional climate and geology (Bailey 1996). By determining moisture conditions, landform (such as the shape of the land surface), and rates of water percolation through the soil, regional climate and geology determine the density, characteristics, and regional importance of wetlands for fish and wildlife. For example, in southern Georgia and Florida, a shallow groundwater table and **karst geology** create shallow depressions, resulting in high densities of depression wetlands; in central North America, receding glaciers created a high density of depressions that hold water for varying amounts of time during the year; and springs in the arid Southwest create important wetland habitat for both fish and wildlife. In turn, the physical environment sets constraints on the **biotic environment**, which is made up of the organisms occurring in a given area.

Hydrology is the most influential physical factor in determining the structure and function of wetland systems, including use by fish and wildlife. The hydrology of a wetland involves the **water budget** and the resulting **hydroperiod**. The water budget considers the set of inputs (precipitation, surface water inflow, groundwater inflow, and tidal-driven inflow) and output (evaporation, transpiration from plants, surface water outflow, and groundwater outflow) of water that determine the temporal patterns of water levels in the wetland. This temporal pattern of water-level fluctuation is referred to as a hydroperiod and varies greatly among different types of wetlands and among years within individual wetlands (figure 1.1). The relative rates of inputs and outputs and underlying geology have a large influence on the physical and chemical conditions of the water in wetland systems. For example, in southeastern depression wetlands, variation in water chemistry among wetlands is related to underlying geology (Newman and Schalles 1990). In turn, variation in plant community structure is related to variation in the water chemistry of these systems. Both water chemistry and vegetative community structure determine the potential use of wetlands by fish and wildlife.

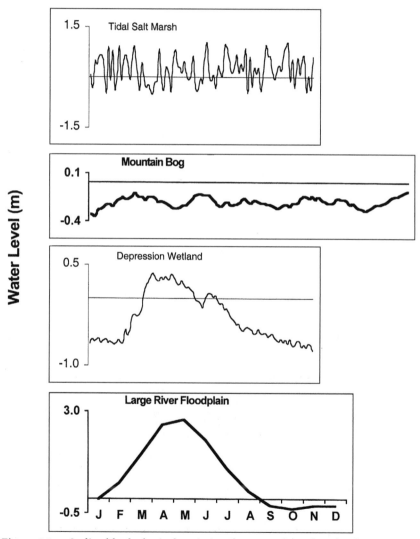

Figure 1.1. Stylized hydrological variation for a number of wetland types. Note the shorter cycles of water level fluctuation in the tidal salt marsh, lack of surface water in the mountain bog, and the annual season patterns of surface water presence and absence in the depression wetland and large river floodplain.

The biotic environment consists of the organisms found in an area. One of the most important components of the biotic environment for fish and wildlife is the vegetation of an area. The plants of an area provide structure and food resources for animal communities, and the diversity of vegetation structure and plant species of wetlands is often positively correlated with the diversity of animals occurring there. In flooded wetlands, general vegetation types include submerged, floating, and emergent aquatic vegetation and shrub and forest canopy layers when present (figure 1.2). Aquatic vegetation provides protection from predators and areas of high production of algae and invertebrate food resources for many aquatic vertebrates. Additionally, aquatic vegetation provides calling sites for male frogs and sites for egg attachment for fish, amphibians, and many invertebrate species. For birds and mammals, aquatic vegetation also provides forage and materials for constructing shelters and nests. For example, muskrats (*Ondatra zibethicus*) consume the roots of aquatic vegetation, and muskrats, alligators, and many species of waterfowl use the leaves and shoots of aquatic vegetation in construction of their nests. The shrub

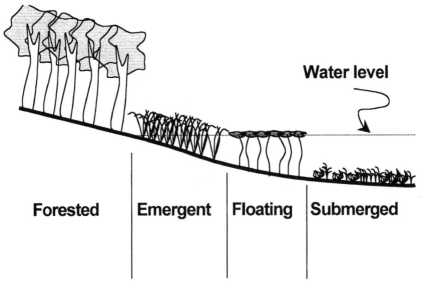

Figure 1.2.   Relationship between topography and the distribution of different plant forms in a saucer-shaped wetland. The mean water level is shown.

and canopy layers provide nesting sites for waterbirds such as cranes (family Gruidae), ibises (family Threskionithidae), and herons (family Ardeidae) and food resources and lodge construction material for beavers (*Castor canadensis*).

Within wetlands, variation in topography and the resulting variation in hydrological conditions influence the distribution of plants. In relatively high-elevation portions of a wetland, the stressful conditions resulting from saturated or flooded soils occur infrequently. In contrast, in lower-elevation areas, saturated or flooded soils may be a permanent condition. This range of conditions favors the development of different plant associations over specific ranges of elevation. In the simplest case of a saucer-shaped depression wetland, bands of different plant associations are found from the center to the outer edges of the wetland (figure 1.2). In low-energy systems, such as isolated wetlands, topology may be relatively stable through time. However, changes in the distribution of plant associations within low-energy systems may result from yearly variation in hydrological conditions driven by annual variation in precipitation. For example, during a series of dry years, soils at the margin of an isolated wetland may not be saturated or flooded, and terrestrial vegetation may encroach into the wetland. In high-energy systems, such as river floodplains, erosion, deposition of sediments and destruction of vegetation during flooding, and meandering of the river channel result in relatively frequent reshaping of topology in these systems. Therefore, the distribution of plant associations is more dynamic in high-energy systems. The soil **seed bank** also plays a role in determining plant community structure and provides plants intolerant of flooded conditions with a mechanism for persisting from one dry period to the next. Viable seeds of many intolerant species persist in flooded wetland soils and germinate when favorable soil conditions (moist but not flooded) return. In some cases, germination and seedling growth can occur only during dry periods, but more mature individuals are tolerant of flooded conditions. For these species, periodic wetland drying is critical for recruitment. This is the case with many emergent grasses and trees associated with wetlands. For example, cypress

(*Taxodium* spp.) required a brief dry period followed by a low-water period for successful seed germination and seedling growth, respectively (Conner and Flynn 1989; DuBarry 1963).

Competition among plants and the action of animals may also structure plant communities in wetlands. For example, cattails (*Typha* spp.), after becoming established in a wetland, will outcompete other species of plants, resulting in monospecific stands of cattails with little or no value to fish and wildlife. Three mammals—nutria (*Myocaster coypus*), muskrats, and beavers—have the ability to extensively alter wetland vegetation. Nutria is an introduced species to North America and, like the native muskrat, feeds on the roots of aquatic vegetation. Nutria destroy large areas of fresh- and salt water marsh vegetation each year and are believed to be responsible for the loss of wetland acreage in some areas (Shaffer et al. 1992). Muskrats play a role in cyclic vegetation patterns often seen in some wetland systems. For example, muskrat destruction and consumption of emergent vegetation in prairie pothole wetlands of the Midwest result in a shift from an emergent vegetative condition to a "lake marsh" condition where submerged aquatic plants are dominant (van der Valk and Davis 1978). Through their cutting of trees and damming of small streams, beavers are responsible for altering the species composition of riparian forest wetlands and converting forested wetlands to emergent marshes (Naiman et al. 1988).

ANIMALS ARE ADAPTED TO WETLAND CONDITIONS

While flooded conditions are environmentally stressful times for plants, aquatic forms of animals find dry or low-water periods stressful. During low-water periods, organisms are crowded into a small volume of water where **anoxic** conditions may develop and waste products may accumulate to toxic levels. When wetlands dry completely, most aquatic forms of animals cannot survive for long. Some animals avoid low-water or dry conditions by migrating to other aquatic habitats, while others produce dormant forms or have larval forms that are present only during wet periods (Williams 1987). Fish may move from floodplains and tidal marshes to adjacent river channels or estuaries during low water. Many insects and amphibians require flooded conditions for reproduction and development of larvae, but adult forms are capable of aerial respiration and do not require flooded conditions. A number of invertebrates reach adulthood during wet periods and produce encysted eggs capable of persisting in wetland soils during dry periods. These encysted eggs hatch when water returns, beginning the life cycle again. A few aquatic animals have evolved physiological mechanisms for dealing with dry conditions; sirens and amphiumas (large aquatic salamanders) of North America and lungfishes of Africa, Australia, and South America burrow into the mud of drying wetlands, where they are able to survive until water returns to the wetlands. Finally, even during high-water periods, environmental conditions can be stressful for many fishes. Stagnant conditions and high rates of decomposition can lead to low dissolved oxygen levels and high temperatures in many wetland systems (figure 1.3). Some fishes have developed tolerances

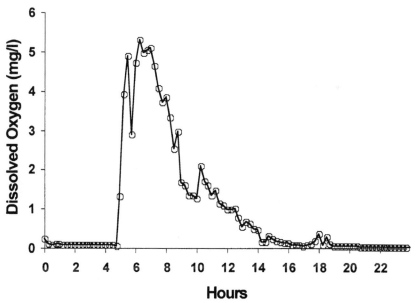

Figure 1.3.   Variation in dissolved oxygen levels at the bottom of a depression wetland in South Carolina during the late spring. Values were taken every 15 minutes over a 24-hour period. Note the almost complete lack of oxygen during some portions of the observation period.

for high temperatures or are physiologically, morphologically, or behaviorally adapted to cope with low dissolved oxygen levels (Kramer 1987).

In wetlands with longer hydroperiods or that dry only during severe drought periods, larger aquatic organisms with longer developmental periods are able to persist. These organisms prey on smaller aquatic forms with shorter developmental periods, often with negative effects on prey populations. For example, the occurrence of predatory fishes in more permanent depression wetlands limits the distribution of some amphibians to depression wetlands with shorter hydroperiods (figure 1.4) (Collins and Wilbur 1979; Snodgrass et al. 2000a, 2000b). Therefore, aquatic forms face a trade-off between developing rapidly to a stage at which they can mi-

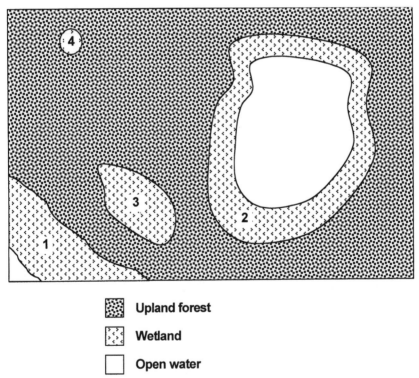

     Upland forest

     Wetland

     Open water

Figure 1.4. Drawing of a hypothetical landscape showing the spatial distribution of wetlands in relationship to upland, open-water, and other wetland habitats. Note the location of wetland 1 and 2 between upland and open-water habitats and the greater isolation of wetland 4 than 3 from other wetland habitats.

grate, metamorphose, or produce drought-resistant forms before shorter hydroperiod wetlands dry or adapting predator avoidance behaviors that allow them to persist in longer hydroperiod wetlands with predator populations (Wellborn et al. 1996). Because adaptation of predator avoidance behaviors often reduces foraging efficiency and developmental rates, it does not work well in shorter hydroperiod wetlands, where slower developing organisms run a greater risk of mortality when wetlands dry. The overall result is the preference of aquatic species for wetlands with a restricted range of hydroperiods.

The seasonal timing of wetland drawdown and refilling is also critical for fish and wildlife that utilize wetlands. Many wetland systems have predictable seasonal (most freshwater wetlands) or daily (tidal wetlands) patterns of low- and high-water conditions (figure 1.1), allowing the adaptation of organisms to these patterns (Junk et al. 1989). Low-water periods produce critical resource pulses for species that feed on aquatic organisms. During low-water periods, aquatic prey become concentrated. Because the seasonality of low-water periods is often correlated with breeding seasons of wading birds, many wading birds depend on this source of concentrated prey to feed their nestlings (Fleming et al. 1994). However, extremely low water conditions or complete drying of wetlands during the breeding season can be detrimental to the nesting success of wood duck (*Aix sponsa*) and wading birds that receive protection from terrestrial predators by nesting over water. In years when wetlands dry, wading birds can experience near 100% nest failure as a result of predation on eggs and nestling by such terrestrial predators as raccoons (Coulter and Bryan 1995). For other organisms that utilize wetlands for spawning and development of aquatic larval and juvenile forms, wetlands must be flooded during the breeding season and remain flooded until larvae or juvenile forms or both reach a development stage at which they can leave the wetland. These organisms include fishes that move from river channels to spawn on inundated floodplains where larvae and juveniles remain and grow, taking advantage of the abundant resources available on the floodplain, and amphibians that move from terrestrial habitats to spawn in depression wetlands when they are flooded in the winter and spring.

WETLANDS ARE PARTS OF LARGER LANDSCAPES

As is suggested by organisms with complex life cycles that use wetlands and adjacent permanent aquatic and terrestrial habitats during different stages, the location of wetlands in the landscape can influence use by fish and wildlife as well as wetland ecosystem structure and function. The landscape position of a wetland refers to its location in relation to other habitat types, such as forests, fields, urban areas, rivers, lakes, and open waters of estuaries and oceans as well as other wetlands. Clearly, the use of wetlands by organisms with complex life cycles depends on the proximity of appropriate upland, open-water, and wetland habitat types. Furthermore, because wetlands are often located in lower areas of the landscape or between upland and open-water habitats, land use in adjacent upland areas often influences inputs of nutrients and pollutants, hydrology, and ultimately use by fish and wildlife (Azous and Horner 2001). The location of wetlands between upland and open-water habitats may buffer the effects of nutrients or other upland pollution sources on open-water habitats. For example, **riparian buffers** (floodplain wetlands) along streams remove the majority of nitrogen and phosphorus in storm runoff and groundwater moving from agricultural fields to streams (Peterjohn and Correll 1984). Landscape position can also affect the amount of groundwater input to freshwater wetlands, influencing both chemical characteristics of the water (Kratz et al. 1997) and hydroperiod.

Metapopulation theory suggests that the spatial relationship of wetlands in landscapes may be important for the persistence of some populations of wetland-associated organisms (Hanski and Simberloff 1997). A metapopulation is a collection of subpopulations occupying individual patches of habitat, with occasional movement of individual among subpopulations that is dependent on their spatial relationships to each other (figure 1.5a). Subpopulations are viewed as occasionally going extinct as a result of stochastic environmental or demographic events and being recolonized by dispersers from other subpopulations. As long as recolonization rates remain high in relation to extinction rates, the metapopulation will persist, although individual habitat patches will experience local extinctions and recolonizations. From a conservation standpoint, the theory suggests that isolation of wetlands through human destruction of wetlands or other natural habitats between wetlands can result in extinction of metapopulations by reducing colonization rates of remaining wetlands (figure 1.5b). It has been suggested that metapopulation theory may be

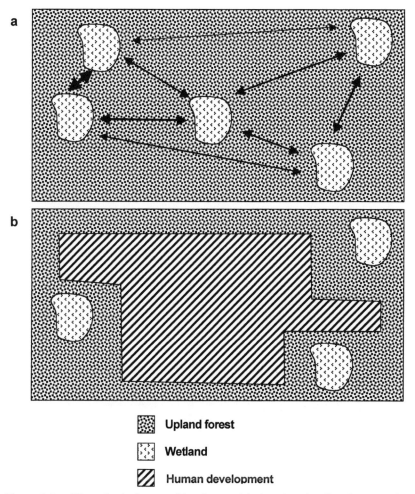

	Upland forest
	Wetland
	Human development

Figure 1.5. Hypothetical natural landscape (a) showing the distribution of wetlands in relationships to each other and the exchange of individuals within a metapopulation of organisms occupying the wetlands. The weight of arrows indicate the amount of movement between subpopulations; greater weight indicates greater amount of exchange. In (b) the landscape has been altered by human development. Note the resulting isolation of subpopulations.

applicable to some wetland types, such as depression wetlands, which exhibit a great deal of interannual variability in hydroperiod lengths. However, it remains unclear whether metapopulation theory is applicable to wetland-associated species (Marsh and Trenham 2001) or to what degree individual populations rely on groups of wetlands in the landscape.

Managing Wetlands for Fish and Wildlife

Fisheries and wildlife biologists may manage wetlands over a series of spatial scales with goals ranging from maintaining or enhancing populations of an individual species to restoring or maintaining the ecological integrity of large systems or collections of wetlands. At smaller scales (within individual wetlands), manipulations of the physical and biological environment may be conducted. Manipulations of the physical environment may include the creation of topographic heterogeneity and the filling of ditches to restore natural hydrological cycles or the control of water levels to promote optimal conditions for a single group of organisms, such as waterfowl. Manipulations of the biotic environment often involve planting of vegetation and removal of introduced plants and animals. Some introduced plants form monotypic stands in wetlands, displacing native vegetation and its associated fauna and altering hydrology and biogeochemical cycles (U.S. Congress 1993). Traditional techniques of controlling introduced plants, such as treatment with herbicides and mechanical removal, have met with limited success. However, **biological control**, the introduction of an insect herbivore specialist from the plant's native range, has been successful at some locations (Blossey 2001). Introduced animals can have devastating effects on wetland systems by preying on native species and destroying natural vegetation communities. These organisms have to be removed by passive or active trapping or in some cases hunting. Care must be taken to avoid harming non–target organisms when attempting to manage introduced animals. For example, nets used to capture introduced fish may also trap and drown native species, such as turtles and snakes.

At larger scales, wildlife and fisheries biologists are faced with managing factors outside individual wetlands that may influence hydrological cycles within wetlands. Dikes and levees have altered exchanges of water between rivers and floodplain wetlands and between estuaries and coastal wetlands. Changes in land use within the drainage basin of depression wetlands can lower groundwater table elevations by increasing **evapotranspiration** and runoff rates, ultimately leading to decreased hydroperiods. Dams operated to reduce flooding or for power generation alter the timing and amount of floodplain inundation for great distances below the dam. While these alterations of wetland landscapes influence hydrological conditions within wetlands, they also serve to isolate wetlands from open-water and terrestrial habitats, further reducing wetland value for fish and wildlife. This isolation may involve barriers to movement between wetlands and other habitats, such as dikes separating floodplains from river channels, or inhospitable land use (such as agricultural fields or urban areas) separating upland forest from wetlands. Fish and wildlife species with complex life cycles requiring both wetlands and deepwater or terrestrial habitats are not likely to persist in such wetlands. Furthermore, wetlands may become "temporally" isolated from other habitats. For example, if fishes that utilize floodplain wetlands for spawning become ripe in the spring, when floodplains are naturally inundated, but water releases from an upstream dam occur mainly during the fall, then the floodplain is not available for the fish when they need it; it has been effectively isolated in time.

Methods for landscape management of wetlands include reconnection of wetlands with open-water habitats, water delivery to mimic natural wetland hydrological cycles, and land use planning. Breaching of dikes and levees that separate floodplain and fringing wetlands of lake, estuary, and ocean from open-water habitats can reestablish natural hydrological cycles and the exchange of organisms, nutrients, and materials between the two systems. These methods are being applied in a number of systems, and early evaluations of their success are positive (Gilmore et al. 1981; Toth et al. 1998). Historical records can be used to reconstruct natural flooding cycles of river floodplains (Richter et al. 1996), and then releases of water from dams can be made to mimic the natural flooding cycle as closely as possible. Again, these methods are also beginning to produce positive results (Molles et al. 1998). Information on the effects of urbanization on wetlands is beginning to emerge, with the potential for incorporation into land planning efforts (Azous and Horner 2001); the success of these efforts remains to be seen. It should be noted that management efforts at larger scales require extensive cooperation among management agencies, scientists, and the public and are more costly and time consuming but will determine the ultimate success of efforts within wetlands. For example, planting of wetland vegetation is unlikely to be successful until external factors affecting hydrological conditions within the wetland are addressed.

Finally, along with an increased focus on management of wetlands for ecological integrity as well as for fish and wildlife has come a need to assess the ecological integrity of wetland systems. The general approach being followed in many systems involves comparisons of physical conditions and biological communities between **reference wetlands** of the appropriate type and wetlands to be assessed or monitored (Rader et al. 2001; Smith et al. 1995). Reference wetlands are relatively pristine systems thought to have high ecological integrity. These methods are currently being used to assess impacts on existing systems, to identify and prioritize systems for conservation efforts, and to monitor the success of wetland restoration and creation projects (see Callaway, this volume).

Conclusion

Decades of research clearly indicate the importance of wetlands for fish and wildlife. The central role of hydrology in controlling fish and wildlife use of wetlands has also been established. The vegetative structure and hydrological requirements of a large number of species have been documented, but information on the majority of species associated with wetlands remains incomplete. This information is likely to continue to accumulate as fisheries and wildlife biologists conduct studies of the natural history of individual species and ecologists continue to test ecological theories in wetland systems. However, we are unlikely to ever have a complete catalog of natural history information for all wetland-associated species. The pattern that emerges from our current knowledge is one in which species have relatively strict requirements for different hydrological regimes so that one group of species is found in wetlands with long hydroperiods, while other species are likely to occur in other wetlands with shorter hydroperiods. For aquatic forms, predator–prey and competitive interactions combine with hydrological variation in shaping the adaptations of many species to wetland life.

It is also clear that wetlands do not exist in isolation but are parts of larger landscapes that include open-water and terrestrial habitats as well as other wetlands. Alteration of habitats adjacent to wetlands or even further removed affects wetland hydrology and the availability of habitats for species requiring wetlands to complete their life cycle. Alterations of habitats adjacent to wetlands can indirectly influence wetland community structure by eliminating organisms with complex life cycles whose aquatic forms often make up a large part of the wetland community. Furthermore, the role of metapopulation dynamics or the use of collections of wetlands by species with complex life cycles remains poorly understood. Because testing of these ideas will require data on demographics and movement of animals among a number of wetlands, this represents a particularly challenging area of future research for fisheries and wildlife biologists.

Fisheries and wildlife biologists face different challenges in managing individual types of systems. In cases where impacts are local (such as introduced species), the continued development and refinement of management techniques is needed, including control of exotics and reestablishment of vegetative diversity and connections with other systems. In other cases (such as floodplain wetlands), the factors controlling hydrological variation are understood, but alterations have been in place for long periods of time, making the identification of natural hydrological regimes difficult. In still other cases (such as depression wetlands), a complete understanding of the factors controlling temporal and spatial hydrological variation is still developing, and quantitative relations between terrestrial land use changes and hydrological alterations are lacking. Research in all these areas is needed and will be vital to the future successes of many management projects.

REFERENCES

Avery, J., and W. Lorio. 1999. *Crawfish production manual.* Louisiana Cooperative Extension Service Publication, no. 2637. Louisiana State University Agricultural Center, Baton Rouge.

Azous, A. L., and R. R. Horner, eds. 2001. *Wetlands and urbanization: Implications for the future.* Boca Raton, Fla.: CRC Press.

Bailey, R. G. 1996. *Ecosystem geography.* New York: Springer-Verlag.

Blossey, B. 2001. Biological control of an invasive wetland plant: Monitoring the impact of beetles introduced to control purple loosestrife. In *Biomonitoring and management of North American freshwater wetlands.* Edited by R. B. Rader, D. P. Batzer, and S. A. Wissinger. New York: Wiley, 451–64.

Boylan, K. D., and D. R. MacLean. 1997. Linking species loss with wetlands loss. *National Wetlands Newsletter* 19(1):13–17.

Collins, J. P., and H. M. Wilbur. 1979. Breeding habits and habitats of the amphibians of the Edwin S. George Reserve, Michigan, with notes on the local distribution of fishes. Occasional Papers of the Museum of Zoology, University of Michigan, no. 686, 1–34.

Conner, W. H., and K. Flynn. 1989. Growth and survival of baldcypress (*Taxodium distichum* [L.] Rich.) planted across a flooding gradient in a Louisiana bottomland forest. *Wetlands* 9:207–17.

Coulter, M. C., and A. L. Bryan Jr. 1995. Factors affecting reproductive success of wood storks (*Mycteria americana*) in east central Georgia. *Auk* 112:237–43.

Cowardin, L. M., V. Carter, F. C. Golet, and E. T. LaRoe. 1979. *Classification of wetlands and deepwater habitats of the United States.* FWS/OBS-79/31. Washington, D.C.: U.S. Department of the Interior, Fish and Wildlife Service.

Dahl, T. E. 1990. *Wetland losses in the United States 1780s to 1980s.* Washington, D.C.: U.S. Department of the Interior, Fish and Wildlife Service.

Diamond, J. 1986. Overview: Laboratory experiments, field experiments, and natural experiments. In *Community Ecology*. Edited by J. Diamond and T. J. Case. New York: Harper & Row, 3–22.

DuBarry, A. P., Jr. 1963. Germination of bottomland tree seeds while immersed in water. *Journal of Forestry* 61:225–26.

Edwards, T. C., Jr. 1989. The Wildlife Society and the Society for Conservation Biology: Strange but unwilling bedfellows. *Wildlife Society Bulletin* 17:340–43.

Environmental Laboratory. 1987. Corps of Engineers wetlands delineation manual. Technical Report Y-87-1. U.S. Army Corps of Engineers Water Experiment Station, Vicksburg, Mississippi.

Fleming, D. M., W. F. Wolff, and D. L. DeAngelis. 1994. Importance of landscape heterogeneity to wood storks in Florida Everglades. *Environmental Management* 18:743–57.

Gilmore, R. G., D. W. Cooke, and C. J. Donohoe. 1981. A comparison of the fish populations and habitat in open and closed salt marsh impoundments in east-central Florida. *Northeast Gulf Science* 5:25–37.

Gosselink, J. G., and E. Maltby. 1990. Wetland losses and gains. In *Wetlands: A threatened landscape*. Edited by M. Williams. Oxford: Basil Blackwell, 296–322.

Hanski, I., and D. Simberloff. 1997. The metapopulation approach, its history, conceptual domain, and application to conservation. In *Metapopulation biology*. Edited by I. A. Hanski and M. E. Gilpin. New York: Academic, 5–26.

Helfield, J. M., and M. L. Diamond. 1997. Use of constructed wetlands for urban stream restoration: A critical analysis. *Environmental Management* 21:329–41.

Junk, W. J., P. B. Bayley, and R. E. Sparks. 1989. The flood-pulse concept in river-floodplain systems. *Canadian Journal of Fisheries and Aquatic Sciences* 106:110–27.

Karr, J. R. 1991. Biological integrity: A long-neglected aspect of water resource management. *Ecological Applications* 1:66–84.

Kramer, D. L. 1987. Dissolved oxygen and fish behavior. *Environmental Biology of Fishes* 18:81–92.

Kratz, T. K., K. E. Webster, C. J. Bower, J. J. Magnuson, and B. J. Benson. 1997. The influence of landscape position on lakes in northern Wisconsin. *Freshwater Biology* 37:209–17.

Marsh, D. M., and P. C. Trenham. 2001. Metapopulation dynamics and amphibian conservation. *Conservation Biology* 15:40–49.

Mather, M. E., D. L. Parrish, R. A. Stein, and R. M. Muth. 1995. Management issues and their relative priority within state fisheries agencies. *Fisheries* 20:14–21.

Meffe, G. K., and C. R. Carroll. 1997. *Principles of conservation biology.* Sunderland, Mass.: Sinauer Associates.

Mitsch, W. J., and J. G. Gosselink. 2000. *Wetlands.* 3rd ed. New York: Wiley.

Molles, M. C., Jr., C. S. Crawford, L. M. Ellis, H. M. Valett, and C. N. Dahm. 1998. Managed flooding for riparian ecosystem restoration. *Bioscience* 48:749–56.

Naiman, R. J., C. A. Johnston, and J. C. Kelley. 1988. Alteration of North American streams by beaver. *Bioscience* 38:753–62.

Newman, M. C., and J. F. Schalles. 1990. The water chemistry of Carolina bays: A regional survey. Archiv für Hydrobiologie 118:147–68.

Peterjohn, W. T., and D. L. Correll. 1984. Nutrient dynamics in an agricultural watershed: Observations on the role of a riparian forest. *Ecology* 65:1466–75.

Rader, B. R., D. P. Batzer, and S. A. Wissinger, eds. 2001. *Bioassessment and management of North American freshwater wetlands.* New York: Wiley.

Richter, B. D., J. V. Baumgartner, J. Powell, and D. P. Braun. 1996. A method for assessing hydrological alteration within ecosystems. *Conservation Biology* 10:1163–74.

Ricklefs, R. E., and Gary L. Miller. 1999. *Ecology.* New York: Freeman.

Ross M. R., and D. K. Loomis. 1999. State management of freshwater fisheries resources: Its organizational structure, funding, and programmatic emphases. *Fisheries* 24:8–14.

Shaffer, G. P., C. E. Sasser, J. G. Gosselink, and M. Rejmanek. 1992. Vegetation dynamics in the emerging Atchafalaya Delta, Louisiana, USA. *Journal of Ecology* 80:677–87.

Smith, R. D., A. Ammann, C. Bartoldus, and M. M. Brinson. 1995. An approach for assessing wetland functions using hydrogeomorphic classification, reference wetlands, and functional indices. Wetlands Research Program Technical Report WRP-DE-9. U.S. Army Corps of Engineers Waterways Experiment Station, Vicksburg, Mississippi.

Smith, R. L. 1996. *Ecology and field biology.* New York: HarperCollins.

Snodgrass, J. W., A. L. Bryan Jr., and J. Burger. 2000a. Development of expectations of larval amphibian assemblage structure in southeastern depression wetlands. *Ecological Applications* 10:1219–29.

Snodgrass, J. W., M. J. Komoroski, A. L. Bryan Jr., and J. Burger. 2000b. Relationships among isolated wetland size, hydroperiod, and amphibian species richness: Implications for wetland regulations. *Conservation Biology* 14:414–19.

Tiner R. W. 1999. *Wetland indicators: A guide to wetland identification, delineation, classification, and mapping.* Boca Raton, Fla.: Lewis.

Toth L. A., S. L. Melvin, D. A. Arrington, and J. Chamberlain. 1998. Hydrologic manipulations of the channelized Kissimmee River—Implications for restoration. *Bioscience* 48:757–64.

U.S. Congress. 1993. Harmful non-indigenous species in the United States. Office of Technology Assessment Report OTA-F-565. Washington, D.C.: U.S. Government Printing Office.

Van der Valk, A. G., and C. B. Davis. 1978. The role of seed banks in the vegetation dynamics of prairie glacial marshes. *Ecology* 59:322–35.

Wellborn, G. A., D. K. Skelly, and E. E. Werner. 1996. Mechanisms creating community structure across a freshwater habitat gradient. *Annual Review of Ecology and Systematics* 27:337–63.

Williams, D. D. 1987. *The ecology of temporary waters.* London: Croom Helm.

SUGGESTED READINGS

Bookhout, T. A. 1994. *Research and management techniques for wildlife and habitats.* 5th ed. Bethesda, Md.: The Wildlife Society.

Cowardin, L. M., V. Carter, F. C. Golet, and E. T. LaRoe. 1979. *Classification of wetlands and deepwater habitats of the United States.* FWS/OBS-79/31. Washington, D.C.: U.S. Department of the Interior, Fish and Wildlife Service.

Karr, J. R. 1991. Biological integrity: A long-neglected aspect of water resource management. *Ecological Applications* 1:66–84.

Junk, W. J., P. B. Bayley, and R. E. Sparks. 1989. The flood-pulse concept in river-floodplain systems. *Canadian Journal of Fisheries and Aquatic Sciences* 106:110–27.

Mitsch, W. J., and J. G. Gosselink. 2000. *Wetlands.* 3rd ed. New York: Wiley.

Tiner R. W. 1999. *Wetland indicators: A guide to wetland identification, delineation, classification, and mapping.* Boca Raton, Fla.: Lewis.

Wellborn, G. A., D. K. Skelly, and E. E. Werner. 1996. Mechanisms creating community structure across a freshwater habitat gradient. *Annual Review of Ecology and Systematics* 27:337–63.

Soils and Sediment

UNDERSTANDING
WETLAND
BIOGEOCHEMISTRY

Stephen Faulkner

Wetlands are important ecosystems that can be distinguished by the presence of water near or above the soil surface, unique soils, and characteristic vegetative communities. Wetlands are inundated or saturated with sufficient duration and frequency to cause changes in the chemical and physical attributes of the area. The changes initiated by the overriding influence of wetland **hydrology** are reflected in the vegetation and soil of the wetland. Plants must possess physiological and/or biochemical adaptations that enable them to survive in saturated soil conditions, resulting in specialized plant communities known as **hydrophytic vegetation**. Since soils take thousands of years to develop, they generally reflect the long-term hydrologic conditions of the site; subsequently, wetland soils look very different from those that develop under drier moisture conditions.

Wetland biogeochemistry is the study of chemical reactions that are mediated by the metabolic activities of biological organisms, especially **microorganisms**. These underlying processes are responsible for the unique ecosystem characteristics of wetlands and the important roles that they can play in the landscape. The complexity of life on earth makes biogeochemistry a difficult subject to simplify, resulting in a poor understanding of the basic concepts and principles. This has created an environment where the protection and management of wetland ecosystems often fall short because of the current inability to incorporate these concepts into regulatory programs.

Wetlands are generally found at the interface between terrestrial and aquatic ecosystems, and this linkage provides important pathways for the flow of energy and materials (figure 2.1). As this material flows into and, in many cases, through the wetland, biogeochemical processes often transform the chemical nature of the material. These transformations play a role in many environmentally important areas and provide the underlying support mechanisms for the significant economic and ecological benefits provided by wetlands, including plant and animal habitat, regulation of water quality and quantity, and a source of food, fiber, and recreation for humans (Mitsch and Gosselink 2000). Costanza et al. (1997) estimated the total value of **ecosystem services** provided by coastal areas and other wetlands worldwide to be $15.5 trillion per year, almost 46% of the total for all global ecosystems.

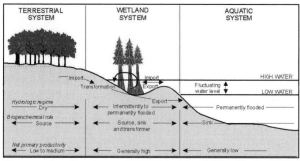

Figure 2.1.   Landscape position of wetlands as an ecotone between terrestrial and aquatic systems (adapted from Mitsch and Gosselink 2000).

This chapter focuses on wetland biogeochemical processes and their impacts on the structure and function of wetland ecosystems. Understanding these processes helps us interpret the myriad of wetland types, their important relationships to other ecosystems in the surrounding landscape, and the important ecological functions they support worldwide.

Scientific Concepts: Oxidation-Reduction Reactions

The surface atmosphere of the earth supports **aerobic respiration** with oxygen functioning as the terminal electron acceptor since it is abundant and an efficient electron acceptor. Oxygen consumption in aerobic soils is easily replenished from the atmosphere through **diffusion** and **mass transfer**. In wetlands, however, saturated soil acts as a barrier since oxygen diffusion is 10,000 times slower through water than air. Oxygen is quickly depleted, creating **anaerobic conditions**. The plants and animals that inhabit wetlands must adapt to these unique soil conditions in order to survive. Once oxygen is depleted, terminal electron acceptors other than oxygen must be used during **anaerobic respiration** (Ponnamperuma 1972).

While **pH** is readily recognized as a measure of proton concentration and infers acid-base status, the concepts of electron chemistry are less well known. The loss of electrons is **oxidation** since, in the early days of chemistry, the known oxidation reactions (such as rust formation) always involved oxygen (Pankow 1991). The gain of electrons is called **reduction** since the addition of negatively charged electrons reduces the overall charge of the element or compound involved. The quantitative measure of the electron availability in a chemical system is defined as **oxidation–reduction** or **redox potential** and is denoted as **Eh** (Ponnamperuma 1972). In practical terms, redox potential measures the tendency of a system to either reduce or oxidize elements or compounds. Redox potential in soils can be measured by inserting an inert platinum electrode into the soil and connecting it to a voltmeter to measure electrical activity that results from electrons transferring from one compound to another.

Redox reactions in wetlands are microbially mediated processes that transfer protons and electrons among redox-active components. Three components are necessary for reduction to occur. First, the soil must be saturated by water and become anaerobic. This forces **facultative aerobes**, microorganisms capable of switching from aerobic to anaerobic respiration, and **obligate anaerobes**, microorganisms that grow only in the absence of oxygen, to use nitrate, manganese, iron, sulfate, and carbon compounds as alternate **electron acceptors** (instead of oxygen) during anaerobic respiration. Electron acceptors are substances that gain electrons during a chemical reaction process. Finally, microbially available organic matter is required to provide electrons and energy. These reactions are dominantly responsible for the development of highly specialized vegetation (hydrophytic) and soils (hydric) characteristic of wetlands. Redox reactions chemically transform compounds and elements and are the basis for **wetland functions** and the ecosystem services they provide. A detailed description of this chemical process is included in appendix A.

These redox processes cause significant changes in the **valence state** of the chemical species used and the overall soil reduction. Reduction of a saturated soil is generally a sequential process governed by the reaction **thermodynamics** (Turner and Patrick 1968). The **redox-active compounds** are shown in table 2.1 along with the **free energy** (ΔG), the approximate redox potential (Eh) of the reaction, and the microbial metabolism involved.

The data in table 2.1 embody an important concept of wetland biogeochemistry: thermodynamic energy yield. The greatest energy yield is derived from aerobic respiration (oxygen reduction) with decreasing energy derived from the suite of alternate electron acceptors used in anaerobic respiration. Those microorganisms deriving the greatest energy will outcompete others and selectively use that electron acceptor, indicating that these processes are driven by microbial metabolism (D'Angelo and Reddy 1999).

Wetland soils undergo significant chemical and biological transformations as alternate electron acceptors are used in a predictable sequence.

Table 2.1. Energy Yield and Redox Potential
of Inorganic Electron Acceptors

REACTION	ΔG	Eh	MICROBIAL GROUP
	Kcal/mol	mV	
Organic matter + $O_2 \rightarrow CO_2 + H_2O$	-686	$>+300$	Obligate aerobes
Organic matter + $NO_3^- \rightarrow N_2 + CO_2$ + H_2O	-649	$+250$	Facultative anaerobes
Organic matter + $MnO_2 \rightarrow Mn^{2+}$ + $CO_2 + H_2O$	-459	$+225$	Facultative anaerobes
Organic matter + $Fe(OH)_3 \rightarrow Fe^{2+}$ + $CO_2 + H_2O$	-100	$+100$	Facultative anaerobes
Organic matter + $SO_4^{2-} \rightarrow S^{2-} + CO_2$ + H_2O	-91	-100	Obligate anaerobes
Organic matter + $CO_2 \rightarrow CH_4 + H_2O$	-88	<-200	Obligate anaerobes

Turner and Patrick (1968) measured the disappearance of oxygen and nitrate and the production of reduced manganese and iron as a function of redox potential (figure 2.2). Oxygen is generally depleted within 24 hours, while the depletion of nitrate takes several days. The plateau in redox potential from day 1 to day 3 reflects the poising of the redox potential until all the nitrate is reduced. The oxidized manganese (manganic, Mn^{4+}) declines as reduced manganese (manganous, Mn^{2+}) increases followed by the presence of reduced iron (ferrous, Fe^{2+}) as both are used as alternate electron acceptors (figure 2.2). These general temporal relationships may be modified by other factors controlling microbial activity, such as temperature, pH, and toxic compounds.

Biogeochemical Cycling

The greater range of redox potentials for wetlands versus nonwetlands is the most important characteristic controlling wetland biogeochemistry.

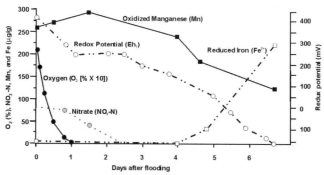

Figure 2.2. Sequential reduction of alternative electron acceptors as a function of flooding duration (after Turner and Patrick 1968).

Wetland ecosystems are dominantly anaerobic and reducing, but they may be aerobic at times or have aerobic zones within the anaerobic environment. Upland systems are dominantly aerobic. The primary aerobic–anaerobic zones are the soil (groundwater wetlands) or water (surface water wetlands) interface with the atmosphere and the area around the roots of wetland plants called the **rhizosphere.** Flood-tolerant plants have developed specialized morphological adaptations that enhance oxygen diffusion to plant roots, and this oxygen diffuses from the roots into the soil, creating an aerobic zone (Armstrong et al. 1994). All aerobic–anaerobic interfaces are characterized by large gradients in redox potential, creating opportunities for redox reactions. Wetlands are the major reducing ecosystem on the landscape and, therefore, have great potential for transforming nutrients and other materials (table 2.1).

The complex interactions of hydrology, vegetative community, soil type, and landscape position result in distinct wetland types that have specific biogeochemical characteristics. The biogeochemistry of different wetland types is affected by inputs to the wetland, intrasystem transformations (or nutrient cycling), and any outflows from the wetland. For example, bogs are *Sphagnum*-dominated peatlands that receive water and nutrients only from precipitation and, therefore, are generally nutrient poor, or **ombrotrophic** (Mitsch and Gosselink 2000). Swamps, on the other hand, may have mineral or organic (peat) soils, are dominated by trees and shrubs, and receive groundwater or surface water inputs high in dissolved minerals yielding a nutrient-rich, or **minerotrophic**, wetland. This section covers the general nutrient cycling processes, while more specific examples are presented in the section "Applied Wetland Biogeochemistry."

CARBON

Photosynthesis and anaerobic respiration dominate carbon transformations in wetlands. Photosynthesis is a redox reaction where oxidized carbon dioxide in the atmosphere is converted to reduced organic carbon, which is then stored as biomass energy in plant tissue (figure 2.3). This plant tissue (organic matter) is composed of long-chained, complex organic molecules, primarily carbohydrates, proteins, and lignin. Respiration oxidizes the reduced organic matter providing energy to heterotrophic organisms. As shown in table 2.1, there are several compounds used as electron acceptors during anaerobic respiration. **Fermentation** is another type of anaerobic

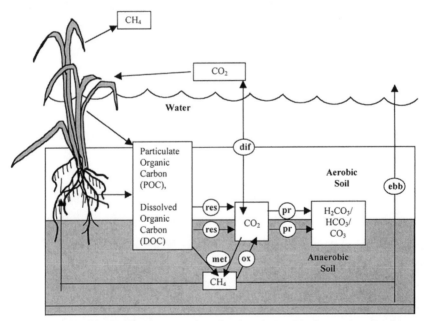

Figure 2.3. Generalized diagram of wetland carbon transformations in the water and soil compartments where dif = diffusion, res = respiration, pr = precipitation, met = methanogenesis, ox = oxidation, and ebb = ebullition.

respiration where one part of the organic molecule is oxidized (loses electrons) and another part is reduced (gains electrons), causing the organic matter to break into smaller organic molecules. Lactic acid, ethanol, and acetate are common end products of fermentation.

The other important carbon transformation in wetlands is **methanogenesis**. A specialized group of bacteria, known as **methanogens**, oxidize carbon dioxide (table 2.1) or the acetate produced by fermentation and produce methane (CH_4) (Schlesinger 1997). Methane is of particular concern because of its role in global warming. This subject is explored further in the section "Applied Wetland Biogeochemistry."

The redox reactions in table 2.1 also demonstrate the importance of carbon in wetlands, which is twofold. First, it is the electron donor for all the reactions. Second, the incomplete oxidation of organic matter under **anaerobiosis** results in the accumulation of organic carbon in wetlands. Moore et al. (1992) reported organic matter half-lives—the time it takes for one-half the material to disappear—of 2.8 years when the oxygen content of wetland soils is at 20%, increasing to 19 years when oxygen content is zero. This accumulation of carbon in wetlands, commonly described as a "carbon sink," has important ramifications. It provides the energy required by the microbial organisms to carry out the transformations of nutrients and pollutants on the landscape and connects wetlands to the global carbon cycle.

NITROGEN

Nitrogen is an essential nutrient required for plant and animal growth. It is also a basic component of proteins, enzymes, and chlorophyll, which allows plants to convert sunlight into chemical energy through photosynthesis. Nitrogen has a complex biogeochemical cycle with multiple biotic and abiotic transformations involving seven valence states ($+5$ to -3). Most of the nitrogen in wetlands is found in the soil (100 to 1,000 grams of nitrogen per square meter) and is predominantly organic nitrogen. Nitrogen stored in the plants is roughly an order of magnitude less than that stored in the soil (Bowden 1987).

Redox reactions in wetland environments transform nitrogen into oxidized and reduced forms (figure 2.4). The dominant pool of organic nitrogen is mineralized to ammonium through ammonification. In most natural wetlands, ammonium is the dominant inorganic form of nitrogen and the primary nitrogen source for plants. In the reduced layer, ammonium is stable and may be **adsorbed** to **sediment exchange sites** or used by both plants and microbes. The juxtaposition of the oxidized and reduced zones in wetlands is important because ammonium is oxidized to nitrate by **nitrifying bacteria** (*Nitrosomonas, Nitrosococcus,* and *Nitrobacter*) through **nitrification** in oxidized zones. Depletion of ammonium in the upper, oxidized layer causes it to diffuse upward in response to the **concentration gradient.**

Since nitrate is the first alternate electron acceptor under anaerobic conditions (table 2.1), it is quickly reduced to either nitrogen gas or nitrous oxide in wetlands through denitrification. Denitrification rates in natural wetlands range from 0.002 to 20 grams of nitrogen per square meter per year (Mitsch et al. 2001). This transformation can rapidly deplete

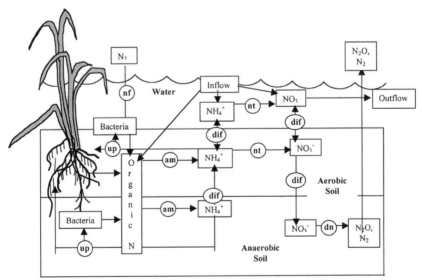

Figure 2.4. Generalized diagram of wetland nitrogen transformations in the water and soil compartments where am = ammonification, nf = nitrogen fixation, dif = diffusion, nt = nitrification, dn = denitrification, up = uptake.

nitrate (see figure 2.2) and creates a concentration gradient for the diffusive flow of nitrate downward in the soil from oxidized to reduced zones. Redox potential, pH, **labile** (microbially available) carbon source, nitrate availability, and temperature control the rate of denitrification (Reddy and Patrick 1984). The processes of ammonification, nitrification, and denitrification dominate wetland nitrogen cycling and potentially process 10 to 190 grams of nitrogen per square meter per year with the major nitrogen flux from natural wetlands coming from denitrification (Bowden 1987; Mitsch et al. 2001). Overall, natural wetlands tend to cycle most of their nitrogen within the system via uptake and mineralization with losses dominated by denitrification.

PHOSPHORUS

Like nitrogen, phosphorus is an essential nutrient for plant growth and a required component of important molecules such as adenosine triphosphate (ATP), the universal energy molecule in organisms. The wetland phosphorus cycle, however, is fundamentally different from the nitrogen cycle. There are no **valency** changes during biotic assimilation of inorganic phosphorus or during decomposition of organic phosphorus by microorganisms. Although phosphorus has no gaseous phase, it has a major geochemical cycle in addition to the biological cycle. In general, phosphorus strongly adsorbs onto soil particles or minerals; therefore, the **sediment–litter compartment** is the major phosphorus pool (>95%) in most natural wetlands, with a much lower plant pool and low concentrations in the overlying water (Richardson and Marshall 1986).

Organic phosphorus; inorganic phosphorus bound in iron, aluminum, and calcium minerals; and orthophosphate are the primary forms of phosphorus in wetland soils (figure 2.5) with organic phosphorus usually the highest (Qualls and Richardson 1995). Mineralization of organic phosphorus converts it to inorganic orthophosphate, which is the form of phosphate used by plants (figure 2.5). The transformations between fixed mineral phosphorus and soluble orthophosphate are controlled by the interaction of redox potential and pH (Holford and Patrick 1979). In acidic soils, orthophosphate is adsorbed to or precipitates with aluminum oxides, iron oxides, or hydroxides, while Ca or Mg compounds dominate at higher pHs. Even though the oxidation state of phosphorus is unaffected by redox reactions, redox potential is important because of iron reduction. Orthophosphate can be released from ferric phosphates under anaerobic conditions as the ferric iron (Fe^{3+}) is reduced to the ferrous form (Fe^{2+}) (Holford and Patrick 1979). Since iron reduction is common in wetland soils, this mechanism is another source of available phosphorus.

SULFUR

Sulfur is an essential nutrient for plant growth and is an integral component of amino acids, proteins, and other cellular components. It is also important in wetlands, as inorganic sulfate is an electron acceptor in anaerobic respiration (see table 2.1). Although there are fewer studies on sulfur

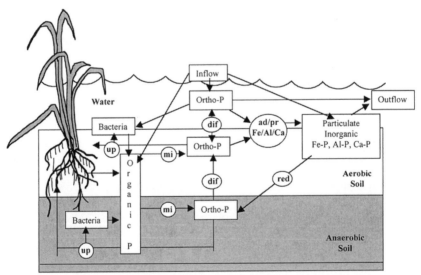

Figure 2.5.   Generalized diagram of wetland phosphorus transformations in the aerobic and anaerobic soil layers where mi = mineralization, ad/pr = adsorption and precipitation reactions with Fe, Al, and Ca, dif = diffusion, red = Fe-reduction, up = uptake.

cycling and retention in wetlands compared to nitrogen and phosphorus, the available data indicate several similarities. The soil is the largest sulfur pool, and it is primarily in the organic form followed by inorganic sulfate, reduced iron sulfides, and a small amount of reduced gases (figure 2.6) (Giblin and Wieder 1992). Despite the small amount of sulfur in the inorganic form, this pool is the most important for cycling, retention, and mobility. Sulfur transformations are also biologically mediated and controlled by the interaction between redox potential and pH.

The organic sulfur in the soil can be converted to sulfate via mineralization (figure 2.6). The sulfate may be used by plants, lost through surface outflow, or reduced through two different mechanisms (figure 2.6). **Assimilatory sulfate reduction** by bacteria reduces SO_4^{2-} and incorporates it into bacterial biomass under oxidizing conditions. Under reducing conditions, **dissimilatory sulfate reduction** transforms sulfate to hydrogen sulfide gas during respiration by obligate anaerobic bacteria (table 2.1 and figure 2.6). Hydrogen sulfide has a very distinctive odor, similar to that of

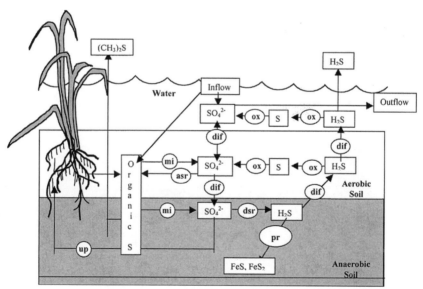

Figure 2.6. Generalized diagram of wetland sulfur transformations in the water and soil compartments where mi = mineralization, ox = oxidation, dif = diffusion, dsr = dissimilatory sulfate reduction, asr = assimilatory sulfate reduction, pr = precipitation, up = uptake.

rotten eggs, and is commonly encountered in wetland soils when such soils are disturbed. It is also toxic to plants, and this toxicity is one of the dominant controls over the types of plants that can grow in wetlands since only a few species have developed the necessary adaptations to deal with this stress (Koch et al. 1990).

Sulfide production in freshwater systems is generally limited by low sulfate levels, but coastal salt marshes have high sulfate-reduction rates. Once formed, hydrogen sulfide can be released to the atmosphere, be reoxidized to inorganic sulfate, or interact with other reduced soil elements (figure 2.6). The generally high levels of reduced ferrous iron in coastal salt marshes react with hydrogen sulfide to form insoluble reduced iron sulfide minerals known as pyrite (an iron ore, FeS, FeS_2). Sulfate reduction and pyrite formation are important mechanisms in salt marshes since they are anaerobic and have plentiful sulfate from seawater and a high soil carbon content. Up to 50% of the carbon decomposition in salt marshes has been attributed to sulfate reduction (Howarth 1993). Sulfide toxicity in wetland soils is lowered by the formation of reduced iron sulfides (Gambrell and Patrick 1978). Without such reductions, the soil would be too toxic for plants and microbes to exist.

Applied Wetland Biogeochemistry

The biogeochemical processes outlined in this chapter are key components of most ecosystem functions and structure of wetlands. The role of biogeochemistry in understanding **wetland structure** and function has many important ramifications in current wetland science. Additionally, the complex interactions among wetland biogeochemical processes, coupled with the wide range of wetland types, provide significant challenges to scientists, regulatory agencies, and natural resource managers. Three specific examples provide insight into these interrelationships and our understanding of what wetlands are and how they work: hydric soils, water quality, and climate change.

HYDRIC SOILS

Wetlands are delineated from nonwetlands by three key criteria: 1) hydrophytic plant community, 2) wetland hydrology, and 3) **hydric soils**. The concept of a hydric soil embodies the overriding influence of water on soil development and encompasses three distinct parts: a definition, database criteria, and field indicators. A hydric soil is defined as one formed under conditions of saturation, flooding, or ponding long enough during the growing season to develop anaerobic conditions in the upper part. The hydric soil definition, however, provides little guidance as to the threshold at which anaerobic conditions develop, what defines the growing season, and how deep the upper part is. This inadequacy is due to the nature of the definition, which must be broad enough to encompass the natural variation of hydric soils.

Database criteria were developed to identify those soil series that meet the definition based on the specific data in the U.S. Department of Agriculture soils database. There are four basic criteria: organic soils (I), mineral soils with high water tables (II), and soils that are frequently ponded (III) or flooded (IV) for long duration (more than seven days). With some exceptions, this approach to identifying hydric soils is useful as a management or planning tool but does not provide a means of identifying hydric soils where it is most important: a given soil at a specific site.

To help resolve this problem, **field indicators** have been developed for the on-site identification of hydric soils (U.S. Department of Agriculture –Natural Resources Conservation Service [USDA-NRCS] 1998). Initially, these indicators have been based on data generated by research soil scientists and the best professional judgment of practicing wetland soil scientists. The indicators are essentially color-related patterns in the soil profile. The basic premise is that these features are a function of **hydromorphic processes** in soils that meet the definition of a hydric soil.

The use of specific macro- and micromorphological features to infer soil wetness is a well-established approach in soil classification (Bouma

1983) and has been the primary basis for selecting specific field indicators (USDA-NRCS 1998). Based on several desirable characteristics, the redox active elements of greatest utility are iron, manganese, and carbon. They are generally abundant in most soils and are actively cycled through redox reactions. Since soil color is derived from compounds containing these three elements, the results of these redox reactions can be observed visually and are strongly correlated with wetland hydrologic regimes and measured soil redox potentials (Faulkner and Patrick 1992).

The field indicators of hydric soils can be grouped broadly into accumulations and depletions of iron, manganese, and carbon and are called **redoximorphic features** (Vepraskas 1992). Oxidized iron and manganese compounds are insoluble and distinctively colored (iron is brown to red, manganese is black). When a soil is flooded or saturated for more than seven days, the oxidized iron and manganese compounds are reduced (see figure 2.2). The reduced iron and manganese are soluble, mobile, and colorless and are readily **translocated** from one area of the soil to another, where they are oxidized back to insoluble, colored precipitates and accumulate in oxidized forms. As wetland soils develop over periods of hundreds to thousands of years, this process results in **depletions** (depleted in iron and manganese) and **concentrations** (accumulations of oxidized iron [brown, red] or manganese [black]). It is this distinctive pattern of soil color that identifies hydric soils and delineates wetlands.

Some soils do not have enough iron and manganese to form redox concentrations and depletions; therefore, the color patterns related to these elements are often absent. The only other element that responds predictably to saturation and anaerobic conditions is carbon. Organic soils are another hydric soil field indicator (USDA-NRCS 1998). They are formed in wetlands when more organic matter is produced by plant growth than can be decomposed under anaerobic conditions. Sandy wetland soils also often lack iron and manganese. Consequently, carbon accumulations and depletions are more reliable field indicators in those soils.

Wetlands and Water Quality

Landscapes are a mixture of land uses and vegetation cover types consisting of various combinations of urban and suburban development, agriculture, forest, and wetland and riparian habitat. In general, when compared to developed areas, natural landscapes generate relatively small amounts of runoff and little water quality impacts. This is primarily because as humans develop land, many of the vegetated areas are replaced with land uses impervious to water penetration. The effect of this development is to raise peak water flows, shorten the time to peak flow, lower water tables, and diminish base flow. Water quality problems result from the increased surface flow of water that dislodges and carries pollutants into downslope ecosystems. The movement of sediment and associated pollutants from terrestrial to aquatic ecosystems is a significant water quality problem in the United States, and the export of these materials is enhanced by anthropogenic disturbances on the landscape.

Non-point-source pollution comes from numerous and widely scattered sources, such as runoff from agricultural areas or municipal storm water, and is the primary cause of impaired water quality in the United States accounting for 65% of the impaired river miles (Carpenter et al. 1998). Excessive sediments and nutrients are responsible for most of the impacts. Sediment loading can destroy fish habitat, increase **biological oxygen demand**, and reduce water storage capacity. In addition, nutrients, metals, and pesticides are often adsorbed onto suspended sediments and deposited in receiving water bodies (Baker 1992). Long-term stream phosphorus concentrations indicate a strong positive correlation with the concentration levels of suspended sediment (Lettenmaier et al. 1991). These pollutants, adsorbed on the suspended sediments, may undergo various biogeochemical transformations including desorption, oxidation–reduction, burial, and biotic uptake (Gambrell and Patrick 1978). Increased concentrations of bioavailable nitrogen and phosphorus create significant water quality problems associated with **eutrophication** of downstream aquatic ecosystems.

Wetlands are uniquely suited to mitigate the negative impacts of non-point-source pollution. Their landscape position and biogeochemical properties (see figure 2.1) give them both the opportunity and the mechanisms to alter pollutant loadings to aquatic ecosystems (Johnston 1991). However, quantifying these capabilities for a specific wetland or class of wetlands requires a more detailed understanding of both the wetland and the chemistry of the pollutant. For example, as previously discussed, nitrogen and phosphorus have different chemical characteristics and different controls on their fate and transport. The reduction of inorganic nitrate to nitrogen or nitrous oxide gas provides a pathway to remove a detrimental nutrient responsible for coastal eutrophication and **hypoxia** (Rabalais and Turner 1996). There is a wide range of denitrification rates across wetland systems (table 2.2) indicating a differential ability specific to the wetland. Similarly, not all restored wetlands have denitrification rates as high as their natural counterparts because of inadequate hydrology or lowered microbial activity (Hunter and Faulkner 2001). Research results suggest loading rates below 20 grams of nitrate per square meter per year will maintain more than 70% removal (Faulkner and Richardson 1989; Mitsch et al. 2001). In contrast to nitrogen, phosphorus has no gaseous outflow and, therefore, will accumulate in wetlands, primarily in the soil compartment (Faulkner and Richardson 1989; Richardson and Marshall, 1986).

In wetlands with mineral soils, phosphorus retention can be predicted by amorphous iron and aluminum oxides (Richardson 1985). These oxides have high surface areas and are chemically reactive as evidenced by their ready dissolution in ammonium oxalate (hence the term oxalate-extractable iron and aluminum). Phosphate coming into the wetland is adsorbed by these oxides and retained in the wetland soil. In wetlands with organic soils and little oxalate-extractable iron and aluminum, phosphate is taken up by plants and converted to the organic form. In these wetlands, phosphorus is retained by the buildup of soil organic matter, effectively burying the organic phosphorus with the organic matter (Craft and Richardson 1998). While initial phosphorus retention by organic accumu-

TABLE 2.2. Nitrogen Loss Rates Reported for Selected Wetland and Riparian Zone Studies	
ECOSYSTEM TYPE	RATES (g N ♦ m^{-2} ♦ yr^{-}1)
Wetlands	
Nutrient-enriched swamp (Florida)	28
Low-nutrient natural wetlands (ave. = 0.19)	0.002–0.34
Nutrient-enriched wetland (ave. = 60)	16–134
Wetlands, Denmark	20–92
River-fed constructed wetlands (Illinois)	171
Treatment wetland, theoretical rate	280
Treatment wetland, based on I/O	801
River-fed constructed wetlands (Ohio)	62–661
Mississippi River diversion at Caernarvon (Louisiana)	10
Riparian systems	
Riparian forest, Chesapeake Bay (Maryland)	4.5–6.0
NO_3 + glucose, buffer zones	80.3
NO_3 + glucose, grass strips	576
Riparian maple swamp (unenriched)	0.5–1.6
Riparian maple swamp (enriched)	2.0–3.6
Restored riparian wetland	6.9
Young hardwood riparian forest	4.3
Alluvial soil	1.5–15.52
Light till	1.0–2.02
Source. Adapted from Mitsch et al. (2001).	

lation or oxalate-extractable iron and aluminum can be as high as 10 grams per square meter per year, this rate is not sustainable since these mechanisms have a finite capacity and, once filled, phosphorus will flow out of the wetland to downstream ecosystems (Richardson et al. 1997). The question, then, is how much phosphorus can be accumulated without causing irreversible changes in wetland structure and function or increasing outflow phosphorus concentrations? While additional research is needed to answer that question for all wetlands, analysis of available data provides some insight. Analysis of outflow phosphorus concentrations as a function of mass loading rate for 126 natural and **constructed wetlands** across the United States indicates a change threshold at a loading rate of one gram of phosphorous per square meter per year (Richardson and Qian 1999). Below this rate, outflow phosphorus concentrations are low and relatively constant, while above this value, outflow phosphorus con- centrations increase significantly with increases in loading rate (figure 2.5). Data from a **eutrophication gradient** in the Florida Everglades support this hypothesis. In areas where phosphorus loading exceeded one gram of phosphorous per square meter per year, there were significant changes in dominant plant species (from sawgrass [*Cladium jamaicense*] to cattail [*Typha domingensis*]) with higher plant productivity, macroinvertebrate diversity, and carbon mineralization rates (Richardson and Qian 1999; Richardson et al. 1997).

It should be evident from this discussion that understanding the biogeochemical controls over nutrient transformations and retention is crucial to predicting both the short- and the long-term impacts of human disturbances in watersheds and wetlands.

Wetlands and Climate Change

Although wetlands only make up approximately 4% of the earth's land area, they store almost 33% of the soil organic matter worldwide, constituting the largest global soil carbon reservoir (Eswaran et al. 1993). High **net primary production** in wetlands, combined with slowed decomposition of organic matter under anaerobic conditions, results in carbon storage rates ranging from 23 to 50 grams of carbon per square meter per year (Schlesinger 1997). This disproportionate amount of carbon storage and the biogeochemistry of organic carbon cycling make wetlands an important component in global climate change, greenhouse gases, and carbon sequestration.

Carbon dioxide and methane account for 80% of the global warming potential of all greenhouse gases (Intergovernmental Panel on Climate

Change 2000); therefore, the release of these two gases from wetlands can have significant impacts on global climate change. When wetlands are drained and soil processes switch from anaerobic to aerobic, organic carbon is more rapidly oxidized to carbon dioxide, and the basic function of the wetland changes from being a carbon sink to a carbon source. This process is magnified by agricultural production (a common use of drained wetlands) and has caused meters of peat to literally vanish from cultivated organic soils (Richardson 1981). The warmer temperatures and lower water tables expected with global warming will likely enhance these processes, particularly in northern peatlands (Freeman et al. 2001).

Wetlands also release methane as an end product of methanogenesis (see table 2.1) and are responsible for 20% to 40% of the annual global atmospheric methane flux (Bartlett and Harriss 1993). Methane is a powerful greenhouse gas with 20 times the warming potential of carbon dioxide; however, methane flux varies among wetland types. Tropical wetlands, with warm soil temperatures augmenting high microbial activity year-round, account for 51% of the total wetland flux, while the lowest emissions come from temperate wetlands (10%) (Bartlett and Harriss 1993). Many temperate wetlands are seasonally inundated during periods of lower soil temperature with lower water tables and aerobic soils in the upper part during warmer months. These conditions not only reduce gross methane production but also allow for significant oxidation, which lowers the net methane emission (Updegraff et al. 2001). Coastal salt marshes generally have low methane emission rates, as the sulfate-reducing bacteria are able to outcompete the methanogenic bacteria for the same substrates (Schönheit et al. 1982).

The complexity of biogeochemical processes, combined with natural variations in wetland soil chemistry and hydrologic regimes, makes it difficult to make general predictions of the response of wetlands to global climate change. Predicted sea-level rise will inundate coastal wetlands and convert them to open water, forever losing land area that currently sequesters carbon. The role of wetlands in the global carbon cycle and their close proximity to rivers and oceans make them an important component of any future climate change.

Conclusion

Wetland biogeochemistry is dominated by transformations from one form to another. The high organic carbon content in wetlands and plant uptake processes enhance the conversion of inorganic elements to organic forms. Redox transformations are microbially mediated through anaerobic respiration and are facilitated by the proximity of oxidizing and reducing conditions in wetlands. Both internal redox cycling and inputs of oxidized species from adjacent terrestrial ecosystems into reducing wetland environments contribute to the magnitude of these chemical transformations.

The inherent variability in natural wetland systems is primarily responsible for the wide range of responses among wetlands. Differences in substrates, hydrologic regime, climate, and loading rates all result in different redox conditions, retention mechanisms, capacities, and reaction rates. These variables also interact with the specific responses, emphasizing the need to identify the pollutants or nutrients of interest and adequately characterize the wetland system under study. Future research will continue to focus on these aspects of wetland biogeochemistry to develop better models for predicting responses to disturbance (especially from climate change and sea-level rise), identify nutrient assimilative capacity and measures of resiliency and integrity, test hypotheses of relationships between wetland structure and function, and identify the connections and impacts to and from other ecosystems at the landscape scale.

REFERENCES

Armstrong, W., R. Brandle, and M. B. Jackson. 1994. Mechanisms of flood tolerance in plants. *Acta Botanica Neerlandica* 43:307–58.

Baker, L. A. 1992. Introduction to nonpoint source pollution in the United States and prospects for wetland use. *Ecological Engineering* 1:1–26.

Bartlett, K., and R. Harriss. 1993. Review and assessment of methane emissions from wetlands. *Chemosphere* 26:261–320.

Bouma, J. 1983. Hydrology and soil genesis of soils with aquic moisture regimes. In *Pedogenesis and soil taxonomy. I. Concepts and interactions.* Edited by L. P. Wilding et al. Amsterdam: Elsevier, 253–81.

Bowden, W. B. 1987. The biogeochemistry of nitrogen in freshwater wetlands. *Biogeochemistry* 4:313–48.

Carpenter, S. R., N. F. Caraco, D. L. Correll, R. W. Howarth, A. N. Sharpley, and V. H. Smith. 1998. Nonpoint pollution of surface waters with phosphorous and nitrogen. *Ecological Applications* 8:559–68.

Costanza, R., et al. 1997. The value of the world's ecosystem services and natural capital. *Nature* 387:253–60.

Craft, C. B., and C. J. Richardson. 1998. Recent and long-term organic soil accretion and nutrient accumulation in the Everglades. *Soil Science Society of America Journal* 62:834–43.

D'Angelo, E. M., and K. R. Reddy. 1999. Regulators of heterotrophic microbial potentials in wetland soils. *Soil Biology and Biochemistry* 31:815–30.

Eswaran, H., E. Van den Berg, and P. Reich. 1993. Organic carbon in soils of the world. *Soil Science Society of America Journal* 57:192–94.

Faulkner, S. P., and W. H. Patrick Jr. 1992. Redox processes and diagnostic wetland soil indicators in bottomland hardwood forests. *Soil Science Society of America Journal* 53:883–90.

Faulkner, S. P., and C. J. Richardson. 1989. Physical and chemical characteristics of freshwater wetland soils. In *Constructed wetlands for wastewater treatment.* Edited by D. Hammer. Chelsea, Mich.: Lewis, 41–71.

Freeman, C., C. D. Evans, D. T. Monteith, B. Reynolds, and N. Fenner. 2001. Export of organic carbon from peat soils. *Nature* 412:785–87.

Gambrell, R. P., and W. H. Patrick Jr. 1978. Chemical and microbiological properties of anaerobic soils and sediments. In *Plant life in anaerobic environments.* Edited by D. D. Hook and R. M. M. Crawford. Ann Arbor, Mich.: Ann Arbor Science Publishers, 375–423.

Giblin, A. E., and R. K. Wieder. 1992. Sulphur cycling in marine and freshwater wetlands. In *Sulphur cycling on the continents.* SCOPE, edited by R. W. Howarth et al., no. 48. New York: Wiley, 85–117.

Holford. I. C. R., and W. H. Patrick Jr. 1979. Effects of reduction and pH changes on phosphate sorption and mobility in an acid soil. *Soil Science Society of America Journal* 43:292–97.

Howarth, R. W. 1993. Microbial processes in salt marsh sediments. In *Aquatic microbiology: An ecological approach.* Edited by T. E. Ford. Oxford: Blackwell, 239–59.

Hunter, R. G., and S. P. Faulkner. 2001. Denitrification potentials in restored and natural bottomland hardwood wetlands. *Soil Science Society of America Journal* 65:1865–72.

Intergovernmental Panel on Climate Change. 2000. Land use, land-use change and forestry. IPCC Special Report. Port Chester, N.Y.: Cambridge University Press.

Johnston, C. A. 1991. Sediment and nutrient retention by freshwater wetlands: Effects on surface water quality. *Critical Reviews in Environmental Control* 21:491–565.

Koch, M. S., I. A. Mendelssohn, and K. L. McKee. 1990. Mechanism for hydrogen sulfide-induced growth limitation in wetland macrophytes. *Limnology and Oceanography* 35:399–408.

Lettenmaier, D. P., E. R. Hooper, C. Wagoner, and K. B. Faris. 1991. Trends in stream water quality in the continental United States, 1978–1987. *Water Resources Research* 27:327–40.

Mitsch, W. J., J. W. Day Jr., J. W. Gilliam, P. M. Groffman, D. L. Hey, G. W. Randall, and N. Wang. 2001. Reducing nitrogen loading to the Gulf of Mexico from the Mississippi River Basin: Strategies to counter a persistent ecological problem. *Bioscience* 51:373–88.

Mitsch, W. J., and J. G. Gosselink. 2000. *Wetlands.* 3rd ed. New York: Van Nostrand Reinhold.

Moore, P. A., Jr., K. R. Reddy, and D. A. Graetz. 1992. The influence of oxygen supply on nutrient transformations in lake sediments. *Journal of Environmental Quality* 21:387–93.

Pankow, J. F. 1991. *Aquatic chemistry concepts.* Chelsea, Mich.: Lewis.

Ponnamperuma, F. N. 1972. The chemistry of submerged soils. *Advances in Agronomy* 24:29–96.

Qualls, R. G., and C. J. Richardson. 1995. Forms of soil phosphorus along a nutrient enrichment gradient in the northern Everglades. *Soil Science* 160:183–98.

Rabalais, N. N., and R. E. Turner. 1996. Hypoxia in the northern Gulf of Mexico: Description, causes, and change. In *Coastal hypoxia: Consequences for living resources and ecosystems.* Coastal and Estuarine Studies 58. Washington, D.C.: American Geophysical Union, 1–36.

Reddy, K. R., and W. H. Patrick Jr. 1984. Nitrogen transformations and loss in flooded soils and sediments. *CRC Critical Reviews in Environmental Control* 13:273–309.

Richardson, C. J. 1981. Pocosins: Ecosystem processes and the influence of man on system response. In *Pocosin wetlands: An integrated analysis of coastal plain freshwater bogs in North Carolina*. Stroudsburg, Pa.: Hutchinson Ross, 3–19.

———. 1985. Mechanisms controlling phosphorus retention capacity in freshwater wetlands. *Science* 228:1424–27.

———. 1989. Freshwater wetlands: Transformers, filters, or sinks? *Freshwater Wetlands and Wildlife* 61:25–46.

Richardson, C. J., and P. E. Marshall. 1986. Processes controlling movement, storage, and export of phosphorus in a fen peatland. *Ecological Monographs* 56:279–302.

Richardson, C. J., and S. S. Qian. 1999. Long-term phosphorus assimilative capacity in freshwater wetlands: A new paradigm for sustaining ecosystem structure and function. *Environmental Science and Technology* 33:1545–51.

Richardson, C. J., S. S. Qian, C. B. Craft, and R. G. Qualls. 1997. Predictive models for phosphorus retention in wetlands. *Wetlands Ecology and Management* 4:159–75.

Schlesinger, W. H. 1997. *Biogeochemistry: An analysis of global change*. 2nd ed. Orlando: Academic.

Schönheit, P., J. P. Krisjansson, and R. K. Thauer. 1982. Kinetic mechanism for the ability of sulfate reducers to out-compete methanogens for acetate. *Archives of Microbiology* 132(3):285–88.

Turner, F. T., and W. H. Patrick Jr. 1968. Chemical changes in waterlogged soils as a result of oxygen depletion. *Transactions of the 9th International Congress of Soil Science* 4:53–64.

Updegraff, K., S. D. Bridgham, and J. Pastor. 2001. Response of CO_2 and CH_4 emissions from peatlands to warming and water table manipulation. *Ecological Applications* 11:311–26.

U.S. Department of Agriculture–Natural Resources Conservation Service. 1998. *Field indicators of hydric soils in the United States, version 4.0*. Edited by G. W. Hurt, P. M. Whited, and R. F. Pringle. Fort Worth, Tex.: U.S. Department of Agriculture, Natural Resources Conservation Service.

Vepraskas, M. J. 1992. Redoximorphic features for identifying aquic conditions. Technical Bulletin 301. North Carolina Agricultural Research Service, North Carolina State University, Raleigh.

SUGGESTED READINGS

Faulkner, S. P., W. H. Patrick Jr., and R. P. Gambrell. 1989. Field techniques for measuring wetland soil parameters. *Soil Science Society of America Journal* 53:883–90.

Faulkner, S. P., and C. J. Richardson. 1989. Physical and chemical characteristics of freshwater wetland soils. In *Constructed wetlands for wastewater treatment*. Edited by D. Hammer. Chelsea, Mich.: Lewis, 41–71.

Gambrell, R. P., and W. H. Patrick Jr. 1978. Chemical and microbiological properties of anaerobic soils and sediments. In *Plant life in anaerobic environments*. Edited by D. D. Hook and R. M. M. Crawford. Ann Arbor, Mich.: Ann Arbor Science Publishers, 375–423.

Kadlec, R. H., and R. L. Knight. 1995. *Treatment wetlands*. Chelsea, Mich.: Lewis.

Mausbach, M. J., and J. L. Richardson. 1994. Biogeochemical processes in hydric soil formation. In *Current topics in wetland biogeochemistry*. Vol. 1. Edited by W. H. Patrick Jr., and J. A. Nyman. Baton Rouge: Louisiana State University, Wetland Biogeochemistry Institute, 68–127.

Mitsch, W. J., and J. G. Gosselink. 2000. *Wetlands*. 3rd ed. New York: Van Nostrand Reinhold.

Ponnamperuma, F. N. 1972. The chemistry of submerged soils. *Advances in Agronomy* 24:29–96.

Reddy, K. R., and E. M. D'Angelo. 1994. Soil processes regulating water quality in wetlands. In *Global wetlands—Old World and New*. Edited by W. M. Mitsch. New York: Elsevier, 309–24.

Richardson, J. R., and M. J. Vepraskas, eds. 2000. *Wetland soils: Their genesis, hydrology, landscape, and separation into hydric and nonhydric soils*. Ann Arbor, Mich.: Ann Arbor Science Publishers.

Stevenson, F. J., and M. A. Cole. 1999. *Cycles of soils: Carbon, nitrogen, phosphorus, sulfur, micronutrients*. New York: Wiley.

3

Restoring Our National Wetlands

UNDERSTANDING
RESTORATION
ECOLOGY

John C. Callaway

Over 50% of the historic wetlands in the United States have been lost in the past two centuries (Dahl 1990), with even greater losses in some states. California has lost 90% of its wetlands, with rates of loss almost as high in some midwestern states (Dahl 1990). Impacts to wetlands have come from drainage for agriculture, port development, urban development, and other reasons. However, over the past three decades, the public has become more aware of the values that wetland ecosystems provide. With this awareness there has been growing interest in the restoration of wetland ecosystems (Casagrande 1997). **Restoration** is the attempt to return a degraded ecosystem to a pristine or "natural" condition. Ecosystems naturally develop, and in restoring wetlands (or other ecosystems), ecologists are trying to speed up the process of natural development. In a sense, we are trying to move a wetland through **successional stages** at a quicker rate than would occur under natural circumstances. In practice, the restoration of wetlands is a complex process involving planning, design, implementation, and monitoring (figure 3.1).

Restoration ecology as a formal discipline is a relatively new field, although humans have been manipulating and "restoring" ecosystems for centuries, if not millennia. One of the first ecologists to attempt to restore an ecosystem using a scientific approach was Aldo Leopold (famous for the book *A Sand County Almanac*). Leopold, John Curtis, Theodore Sperry, and other ecologists initiated the restoration of the Curtis Prairie at the University of Wisconsin Arboretum in 1934, and Leopold's insight into the process has motivated many restoration ecologists. Leopold (1966) recommends "to keep every cog and wheel is the first precaution of intelligent tinkering," and he asks, "Have we learned this first principle of conservation: to preserve all of the parts of the land mechanism?" In the 1970s, restoration ecologists began to restore ecosystems more systematically, and since then there has been increased interest in restoration ecology, including the establishment of the journal *Restoration Ecology* in 1993. With the growing interest in global biodiversity (Tilman 2000), there also is the realization that restoration is a critical management tool for protecting biodiversity.

Although restoration attempts have been made for many different types of ecosystems, wetlands have been at the center of attention for restoration ecology for the past three decades. This is due in large part to the substantial loss of wetlands noted previously but also to the fact that wetlands are uniquely protected habitats. Federal regulations as dictated by the guidelines for Section 404 of the Clean Water Act (see Hague, this volume) call for the **mitigation** of impacts to wetland habitats. The three-step process of mitigation requires that projects first attempt to avoid any inappropriate impacts to wetland habitats; second, minimize any unavoidable impacts; and third, replace the lost wetland area and function through

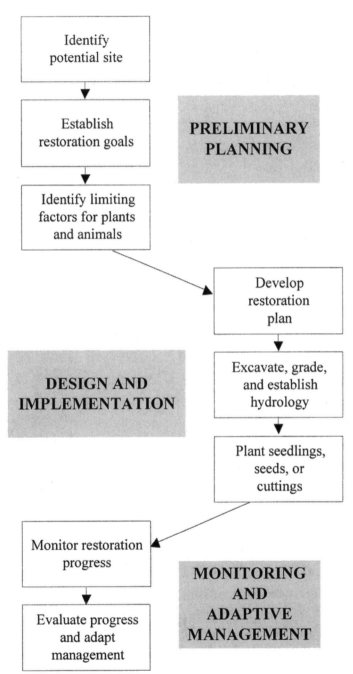

Figure 3.1. Steps in the process of wetland restoration.

restoration, creation, enhancement, and, in exceptional cases, preservation of wetland habitats (National Research Council 2001). The requirement for mitigation of wetlands through restoration has been one of the key reasons why so much of the focus of restoration has been on wetland ecosystems. In addition, interest in the restoration of endangered species habitat has been a driving force in the development of restoration ecology (Falk et al. 1996); wetland ecosystems are host to a large number of endangered species, so this also has focused the attention of restoration ecologists on wetland ecosystems. Substantial information is available for some of the larger restoration projects that have been completed or are currently being planned across the country (table 3.1). In addition, many useful case studies have been compiled in other volumes (see Cairns 1995; Kusler and Kentula 1990).

In discussing restoration ecology, it should be pointed out that restoration means many different things to different people. Part of the difficulty in interpreting restoration is that the definition calls for the return of a degraded system to pristine or natural condition. But how do we define "natural," especially if we are working in a system that has been degraded for an extended period? Does this mean returning it to a condition prior to any impacts from human manipulations? Or does it mean returning it to conditions that existed under Native American (or other indigenous peoples) habitation? Native Americans clearly manipulated ecosystems, although typically their manipulations were similar to natural disturbances that are common in most ecosystems. Or how do we define natural if no records exist of historic conditions at a particular location? Furthermore, some ecosystems may be so highly degraded or the adjacent ecosystems so highly modified that it would be impossible to determine what the pristine condition of the ecosystem may have been. Given these concerns, it should be clear that there is no single definition of natural conditions that will satisfy all people interested in restoring ecosystems.

In addition to the concerns related to the definition of natural conditions, the term "restoration" is used with a range of connotations, from specifically referring to manipulations that return an ecosystem to its prior

Table 3.1. Examples of some large-scale and well-studied wetland restoration projects (see citations for case studies of these projects and for other information).

LOCATION INFORMATION	PREDOMINANT HABITAT TYPES	FURTHER
Delaware Bay, Del.	Salt marsh	Weinstein et al. (2001)
Florida Everglades, Fla. (2001)	Freshwater wetlands www.evergladesplan.org	Chimney and Goforth (2001)
Kissimmee River, Fla.	Freshwater wetlands Riparian scrub	Toth et al. (1998) Middleton (1999)
Mississippi River Delta, La.	Salt marsh Freshwater tidal wetlands	Steyer and Llewellyn (2000) www.lacoast.gov
Olentangy River, Ohio	Freshwater riverine wetlands	Mitsch et al. (1998)
Des Plaines River, Ill. (1994)	Freshwater riverine wetlands	Sanville and Mitsch (1994)
Great Plains, N.D., S.D., Minn., Iowa	Prairie potholes	Galatowitsch and van der Valk (1994)
Sacramento– San Joaquin Delta and San Francisco Bay, Calif.	Riparian wetlands Freshwater tidal wetlands Salt marsh	Alpert et al. (1999) calfed.ca.gov
Tijuana Estuary, Calif.	Salt marsh	Zedler et al. (2001)

condition to the very general sense of any manipulation of an ecosystem that improves overall **ecosystem functions**. In the general sense, restoration may refer to activities from enhancement to habitat creation. In all cases, we are interested in improving ecosystems functions, and the different terms refer to the starting and ending points for a particular ecosystem that is being manipulated. In addition, some ecologists use the term **remediation** to refer to the restoration of highly degraded areas. These are areas that are impacted by toxic materials, and the goal of remediation efforts is not to restore pristine conditions but rather to clean up an area so that it could be used again and to prevent further deterioration of the area.

Regardless of the starting point, the overall goal of restoration ecology is the establishment of a functioning, self-sustainable ecosystem that is integrated into the landscape (National Research Council 1992). The importance of this definition is that we restore ecosystems that do the following:

1. *Provide ecosystem functions.* Ecologists typically group wetland ecosystem functions into three areas (National Research Council 1995): hydrologic processes, biogeochemical processes, and habitat and food web support (table 3.2).
2. *Need minimal human intervention in order to maintain their functions over time.* Restored ecosystems are not gardens. Concerns for the sustainability of restored ecosystems include physical stability (changes

---

Table 3.2. Ecosystem Functions Provided byWetlands, Including Both Natural and Restored Wetlands.

Hydrology 　　Water storage and flood reduction 　　Maintenance of groundwater table 　　Maintenance of surface water levels
Biogeochemistry 　　Maintenance of overall water quality 　　Cycling and transformation of nutrients and other elements 　　Retention of sediment 　　Retention and transformation of dissolved substances and pollutants 　　Accumulation of carbon/peat
Habitat and food web support 　　Maintenance of primary productivity 　　Maintenance of characteristic plant communities that serve as habitat 　　Sustaining anadromous fish and other wetland-dependent aquatic species 　　Maintenance of biodiversity

in hydrology, sedimentation, and erosion), the ability of populations to reproduce over time, the establishment of disturbance regimes (where appropriate), and the ability of species to reestablish following disturbance events. In addition, restoration efforts must address the ongoing causes of degradation (such as exotic species or impacts within the watershed that led to wetland impacts) in order to establish ecosystems that will be sustained over the long term.

3. *Are a part of the larger landscape.* The landscape position of wetlands must be considered; for example, vernal pool wetlands occur only in particular locations within the landscape, as do bottomland hardwood wetlands. Furthermore, most wetlands are transitional habitats that provide a connection between adjacent aquatic and upland habitats. Restoration projects must provide for connections to adjacent habitats so that organisms that use multiple habitats can move easily from one habitat to another. Some types of wetlands also provide for habitat connectivity on a larger scale, such as riparian wetlands that provide for a band of habitat that allows for migration throughout a watershed. Without these connections to the larger landscape, restored wetlands will not provide many of the important ecosystem functions that natural wetlands provide.

A complication for restoration ecology is that the objectives for restoration are often mixed between both ecological objectives and public policy objectives. In an ideal world, these two objectives would be the same; however, for many restoration projects, this is not the case. As noted previously, many wetlands are restored as part of mitigation agreements that allow for the degradation of natural, functioning wetlands. As part of a mitigation agreement, specific success criteria are usually established, typically focusing on the acreage of wetland habitat to be restored and a few simple measures of plant establishment and animal use, with little consideration of function or sustainability. This is commonly done because the evaluation of function is difficult and time consuming. However, because of this, restored wetlands may be determined to be "successes" in terms of mitigation criteria while not meeting any of the ecological criteria that we would associate with a natural, functioning wetland.

Theoretical Concepts for Restoration Ecology

Restoration has been called the "acid test" of ecology because of the substantial challenge that it represents (Bradshaw 1985). In order to restore an ecosystem, we need to first understand all the components of the ecosystem. Just as Aldo Leopold pointed out, we need to preserve all the parts of the ecosystem. Furthermore, beyond just including the components and understanding them, we also must understand the interactions between these components. For example, how do soil nutrient conditions affect competition between plant species? Or how do shifts in the frequency of flooding affect fish access to a wetland and their feeding opportunities, and how are these changes reflected across the food web? In many cases, some of these connections may not be obvious, and it is only when we try to re-create an ecosystem that we understand the significance of these interactions. The restoration of ecosystems is ecology taken to the extreme: we use all the concepts of ecology and all the specific knowledge of an ecosystem in an attempt to guide the reconstruction of the ecosystem.

A special issue of *Restoration Ecology* gives an overview of the conceptual issues that contribute to restoration ecology (for an introduction to this special issue, see Allen et al. 1997), including reviews from the perspective of genetics, populations, succession theory, communities, ecosystems, landscapes, and more. This series of articles provides a very good overview of the theoretical framework of restoration ecology. These ecological theories can give us insight into the development of restoration sites, but, just as important, monitoring and experimentation with restoration projects also give us an outstanding opportunity to test ecological theories and extend our basic understanding of ecological processes (Allen et al. 1997). Zedler (2000) reviews the relevance of existing ecological theories specific to wetland restoration ecology, including theories of island biogeography, niche, population, and trophic interactions, and points out that succession theory is one of the critical issues that must be addressed in trying to understand and predict how restored wetlands might develop over time. Much of the focus in evaluating the application of ecological theory to restoration has been to identify the factors that limit the ecosystem development of restored sites.

LIMITING FACTORS

Water is the crucial factor determining wetland development (Mitsch and Gosselink 2000), and for restoration ecologists it is clear that in order to restore a functioning wetland, we must establish the proper **hydrology**. Wetland hydrology describes water depth, frequency of flooding, and the duration of flooding for a particular wetland. Wetlands may range in hydrology from areas that are flooded for extended periods (such as cypress swamps) to those where the soil is rarely flooded but remains saturated for extended periods (such as vernal pools). The length of time that a wetland is flooded or saturated has a critical effect on many soil processes, including soil development, oxygen availability (long periods of inundation will lead to the development of **anaerobic conditions**), nutrient cycling, and other variables (see Faulkner, this volume). In turn, these soil processes will affect growing conditions for plants and shape the characteristics of a wetland (figure 3.2). The water level or period of inundation at a particular wetland may vary because of tides, river stage, groundwater, precipitation, surface water runoff, or other hydrologic factors.

The hydrology of a specific wetland will be determined in large part by its position within the landscape. Wetlands occupy a unique location within the landscape—at the interface between terrestrial and aquatic ecosystems. Many wetlands tend to be found in the lower part of **watersheds**, where water collects and accumulates. Restoration activities must be mindful of this landscape constraint. Bottomland hardwood wetlands can be restored only in bottomlands, and tidal wetlands must be in an area that will be under the influence of tides. Although this seems obvious,

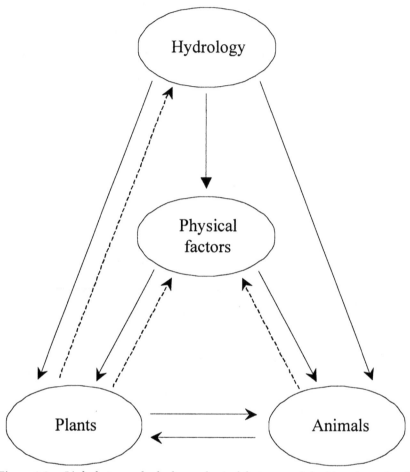

Figure 3.2.   Links between hydrology, physical factors, and the biota in wetlands. Physical factors include oxygen availability, soil wetness, soil nutrient cycling, soil and water chemistry, and other factors. Primary biological feedbacks are indicated with dashed lines.

many attempts at habitat creation have failed because landscape issues were not considered; uninformed restoration ecologists thought that they could create a habitat anywhere they wanted. On a finer scale, creek shape and size are also important to the development of restored wetlands. Most natural wetlands are characterized by networks of many creeks, with a range of sizes and shapes, while restored wetlands tend to have fewer creeks that are simpler and straighter than those in natural wetlands. Hydrologic differences will also affect the biological development of restored wetlands; typically, restored wetlands have only large creeks and lack the small, shallow creeks that are the major access point for aquatic organisms when water levels are high (such as during floodwater or during high tide). Since hydrology is the driving force for wetland development, it must be the first consideration in designing the particular conditions for a restored wetland. Restoration planning should evaluate how the hydrology of a site has been altered and what opportunities and constraints exist for restoring the former hydrologic conditions.

In addition to hydrology, a second important physical consideration is the soil at a restoration site. Wetlands have unique **hydric soils** that typically are dominated by fine soil particles, have relatively high organic content, and experience periods of anaerobic conditions (Mitsch and Gosselink 2000). In order to restore a fully functioning wetland, appropriate soil conditions must be established at the restoration site; however, many restoration projects are established in areas with degraded soils or in areas that were formerly uplands with entirely different soil characteristics. On the other hand, adding too much fertilizer to restoration projects may invite undesirable weeds rather than the target native species. Naturally, soils take centuries to millennia to develop (Jenny 1941) because of the slow processes of organic matter accumulation, mineral transformation, leaching, and so on. Early restoration projects did not consider soil conditions directly; however, it has become clear that this is a critical issue for improving the progress of wetland restoration projects.

The physical conditions at a restoration site, in particular hydrology and soil, will set the limits for the biological development of the restored wetland (figure 3.2). **Succession** is the natural development of ecosystems over time, including shifts in plant species and changes in the physical environment (such as shading and soil development). The study of succession has been a central component of ecology for the past century. In the past few decades, it has become apparent that natural disturbance regimes, including cycles of floods or fires, are an integral part of the succession processes of most ecosystems. In general, as communities go through the process of succession, they develop from relatively simple systems to more complex systems, and natural disturbances serve to set communities back in this continuum. Mitsch and Gosselink (2000) outline in detail the application of general succession theory to wetland development.

As noted previously, one of the goals of restoration ecology is to move wetlands through the successional process as quickly as possible in order to reach a desired community. Under natural conditions, ecosystems may take centuries or more to develop deep, organic-rich soils and other conditions characteristic of late successional stages (Dobson et al. 1997). In order to jump successional stages or move more quickly through the succession process, it is necessary to identify the factors that limit the development of a restored wetland (Dobson et al. 1997; Zedler 2000). Limiting factors may be dispersal of seed or other propagules, nutrient availability, or other soil conditions. However, limiting factors in restored wetlands may be different from those found in natural settings because of alterations to hydrology or other physical factors (such as soil wetness, salinity, or nutrient availability). The succession concept is very useful in describing the general large-scale and long-term patterns of development of wetlands; however, because of the unique factors that may constrain the development of restored wetlands (such as hydrology, soil problems, fragmentation of habitat, and so on), it cannot predict how a specific restoration site may develop in the short term (Zedler 2000). In addition to identifying limiting factors, restoration ecologists must determine appropriate restoration methods that can be utilized to reduce or remove these limits. Although the wealth of knowledge is growing about particular wetland ecosystems such as tidal wetlands (Zedler 2001), riverine systems (Middleton 1999), prairie wetlands (Galatowitsch and van der Valk 1994), and freshwater wetlands (Hammer 1997), the implementation of restoration techniques remains a challenge.

Furthermore, it should be clear that individual restorations sites are complex and unique. While basic ecological theory offers insight into the general development of restored ecosystems, it does not give specific insight into the immediate problems faced by restoration ecologists. For example, existing theory does not help explain what species should be reintroduced into a particular location or how to restore the specific hydrology of an impacted wetland (Zedler 2000). In order to develop specific plans for the restoration of a particular wetland, we need to understand the unique physical constraints at a site and the biology and life history of the target species for that wetland.

MONITORING GOALS AND METHODS

As indicated previously, one of the critical issues for restoration ecology is evaluating the progress of restoration projects. In the past, the focus in evaluating restoration projects has been on identifying success or failure. However, the evaluation of projects in terms of success or failure depends substantially on the criteria that are used for this determination, whether this is in terms of ecological or mitigation criteria (Zedler and Callaway 2000). In the past, the criteria used in evaluating many restoration projects have been minimal. Although restored wetlands may be identified as successful projects, many of these projects do not meet the basic goal of establishing functioning, self-sustainable ecosystems. Because of this problem, Zedler and Callaway (2000) propose that the evaluation of restored wetlands should focus on the progress that a particular project has made toward specific criteria (such as a comparison over time of plant productivity at a restored wetland and a companion reference site) rather than on success or failure. In addition, many early restoration projects had goals that were so vague that it is extremely difficult to evaluate these projects at all. Over the past decade, there has been a substantial effort to improve the goals and criteria for restoration projects, first by establishing clear goals before projects are implemented. These goals should include quantifiable criteria that can be evaluated objectively. Given the wide range of restoration activities, goals do not need to be uniform across restoration projects (Ehrenfeld 2000).

In order to evaluate restoration goals, sites must be monitored over the course of their development. In addition to problems with establishing goals, early restoration attempts frequently lacked sufficient monitoring and evaluation (many had no monitoring at all), and little was learned from these early efforts. These monitoring efforts have improved substantially but still need additional improvements. Many monitoring plans now use an **adaptive management** approach (Thom 2000), using monitoring information to iteratively revise management decisions and goals (figure 3.1). For example, the restoration of wetlands in the Mississippi River delta is one of the largest restoration projects in the country. Substantial efforts have been undertaken to incorporate the dynamic nature of coastal Louisiana into the management program; each restoration project is seen as a way to add to existing knowledge and to improve subsequent management decisions (Steyer and Llewellyn 2000). Further improvement has been to include simultaneous monitoring of natural wetlands in order to compare the development of a restored wetland to a reference ecosystem. The reference wetland should be similar in terms of landscape position, hydrology, species composition, and other physical and biological characteristics. In theory, the restored wetland should become more similar to the reference wetland over time, ideally following some sort of predictable path or "trajectory" of development. The trajectory approach is based on the succession of natural ecosystems, although restored wetlands may not always follow these trajectories because of limitations to development or because of unusual starting conditions (Zedler and Callaway 1999).

OTHER SOCIETAL ISSUES

In addition to the scientific challenges of restoring wetlands, there also are many social challenges. There is substantial interest in restoration from the perspective of human involvement (Casagrande 1997), and it has been pointed out that restoration offers an outstanding opportunity for people to experience natural ecosystems. By involving students, volunteers, and others in restoration projects, people have the chance to create a stronger bond with their natural surroundings and to increase their understanding of the processes that maintain natural and restored ecosystems. Some pro-

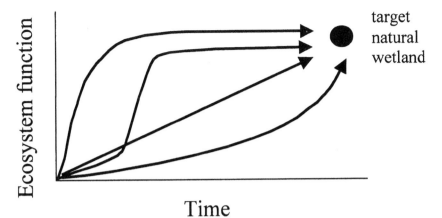

Figure 3.3.   Hypothetical trajectories for the development of ecosystem functions for restored wetlands (modified from Kentula et al. 1992). Each trajectory indicates a potential pathway for the development of ecosystem processes over time.

jects may not allow for the opportunity of public involvement in implementation, but the public should be included in the planning phase. Where possible, restoration ecologists also should include public involvement in restoration implementation. Without public support for restoration, we will fight a losing battle in protecting and restoring wetlands and other ecosystems. Furthermore, there are many societal benefits from restoration because of the functions that wetlands provide, and many cities and states have developed stewardship programs to encourage the restoration and management of wetlands (Cairns 1995; Casagrande 1997).

Current Research in the Field

TRAJECTORIES

One of the areas of interest in the recent restoration research has been to track changes in the functioning of restored wetlands over time. This interest focuses on improving both our understanding of the factors that limit the development of restored wetlands and our ability to predict how future restored wetlands will develop. The trajectory approach has been proposed to describe this process (figure 3.3; Dobson et al. 1997; Kentula et al. 1992); however, few restoration projects have data over a long enough time period to evaluate their development in this manner (Zedler and Callaway 1999). Based on current data, there is only mixed support for the applicability of trajectories in predicting the development of restored wetlands.

Simenstad and Thom (1996) used this approach to assess an estuarine wetland in Washington State's Puget Sound, the Gog-Le-Hi-Te Wetland, over a seven-year period. Multiple ecosystem attributes were measured (including soil, sediment, productivity, invertebrates, fish, and birds) with mixed results. Of the 16 attributes, only species richness of invertebrates and fish, the density of fish, and bird usage indicated trajectories toward **functional equivalency** with a natural, reference wetland. Zedler and Callaway (1999) evaluated plant and soil characteristics at a created salt marsh in San Diego Bay, California, and found very little support for the development of trajectories. Only soil nitrogen concentrations showed a continual increase at the created wetland (relative to a reference site), and at the current rate of increase, it would take over 40 years for the mitigation site to equal conditions at the nearby reference wetland. Craft et al. (1999) used a 25-year record of a restored salt marsh in North Carolina and found that vegetation in the restored marsh reached equivalency with the reference site in less than five years. Invertebrates reached equivalency in five to 10 years, while soil parameters were not equivalent after 25 years. It is likely that some of the differences in trajectories among these sites may be due to the level of degradation at the original restoration site; more degraded sites will take longer to develop and are less predictable in their development.

The use of trajectories in evaluating restoration sites is useful and could help identify limiting factors in the development of restored ecosystems, but the application of this method to predict the development of sites under mitigation policy could be problematic (Zedler and Callaway 1999). Given the lack of evidence to date, we should not base mitigation policy on the belief that a restored wetland will develop along a predictable trajectory. If we allow for permitted impacts to natural wetlands with the expectation that restored wetlands will develop along a trajectory and replace the lost ecosystem functions in the future, we are likely to end up with a net loss in function because the restored wetland may never function as well as the wetlands that were impacted.

PHYSICAL FACTORS

In addition to the evaluation and monitoring of the development of restored wetlands, there is a real need to determine the factors that limit the development of particular restoration sites (Simenstad and Thom 1996). One of the obvious considerations for restored wetlands is hydrology. At minimally disturbed sites, the restoration of appropriate hydrology should be relatively easy. However, there may be significant limits to what can be done at severely degraded sites or at sites with significant logistic constraints for restoring complete wetland hydrology (such as sites that may be affected by adjacent land use or isolated from a hydrologic source by a road or other impediment). The focus of hydrologic restoration should be to restore naturally variable hydrologic conditions (National Research Council 2001). This should be done through the reestablishment of natural hydrology rather than through engineered structures or controls. The focus of most restoration has been on surface water; however, one of the real challenges of restoration is to reestablish the hydrologic link between groundwater and surface water in restored wetlands.

Problems related to soil development at restoration sites also have received extensive research. Both Langis et al. (1991) and Lindau and Hossner (1981) showed that improper soil texture (determined by the percentage of sand, silt, and clay in the soil) is a common problem at restoration sites, leading to significant impacts on soil nutrients and other properties, with eventual impacts on plant growth. Typically, soil texture problems are due to the fact that restored or created wetlands may be established on sandy dredge spoil material or in areas that are excavated from coarse soil texture. Improper soil texture is likely to lead to many further problems at restored wetlands (for a review of these problems, see Callaway 2001). Soils with coarse texture are likely to lose nutrients to leaching. In addition, coarse texture promotes decomposition and reduces the accumulation of soil organic matter, which is the major pool for nutrients in many wetlands soils. Together, these two factors lead to low levels of soil nutrients (especially nitrogen), and low levels of soil nitrogen and other nutrients will significantly affect plant growth. Recent attempts to transplant substrate from impacted natural wetlands have been successful not only because this creates the proper physical conditions for plant growth (correct soil texture, organic content, and nutrient levels) but also because this promotes the establishment of native plant species by importing native species roots, rhizomes, seeds, and other types of plant propagules (Brown and Bedford 1997).

In addition to work on soil conditions, there has been substantial interest in the use of restored and created wetlands to improve water quality through the removal of nutrients, pollutants, and sediments. Much of the interest in water quality has focused on wastewater treatment wetlands, and these are primarily created wetlands that are designed specifically for this function (Kadlec and Knight 1996). Because of unique soil biogeochemistry features in wetlands (see Faulkner, this volume), they can be very effective in removing nutrients and pollutants (especially nitrogen); however, the focus on this function can reduce the value of other ecosystem functions at these wetlands. Given the potential accumulation of pollutants in treatment wetlands, these wetlands usually have little value in regard to wildlife or other habitat functions; however, they do provide benefits for society (such as transformation of nutrients and pollutants) as well as opportunities for education, research, and other human uses.

PLANTS

The greatest focus for most wetland restoration projects has been on plants. Since plants provide the physical structure that creates the habitat, early efforts at restoration focused entirely on establishing vegetation. However, as research with hydrology, soils, and other factors has made clear, simply planting the species of choice is not always enough. We need to remove the physical constraints on plant establishment and growth and create conditions that favor native plants (such as proper hydrology, soil nutrients, and other soil conditions). In addition, we need to consider the biological processes that may affect establishment and growth, such as limits to pollination and seed dispersal, germination needs, competition, and other processes.

Furthermore, when we establish vegetation at a restoration site, just establishing plant cover is not enough; species composition and the canopy characteristics of the vegetation are also critical. Early restoration efforts focused simply on establishing plant cover, often including only a few easy-to-grow species, with little concern for the diversity of species in the restored wetland. Although plants may have become established, the restored wetland was not similar to natural wetlands. For example, it is easy to grow cattails in restored freshwater wetlands, but this species alone will not provide the function of a diverse, natural freshwater wetland. If we allow restoration projects to focus just on the species that are easy to grow, we will end up with poorly functioning, species-poor wetlands, resulting in the loss of function and local biodiversity.

Recently, there has been interest in evaluating the link between plant species diversity and ecosystem function in restored salt marshes (Zedler et al. 2001). Increasing native species richness in restored salt marshes increased important functions within the wetland. Significantly more nitrogen accumulated in areas with greater plant species diversity, indicating that the establishment of diverse plant communities may be one way to speed up development of restored wetlands (Zedler et al. 2001). These links remain to be evaluated in other wetland ecosystems.

A related area of interest concerning plant species at restored wetlands is the issue of restoration "self-design." This relates to the question of whether plant species need to be introduced into a restoration site (a "designer" wetland) or whether the wetland should be allowed to develop on their own (a "self-design" wetland.) In order to evaluate this issue, Mitsch et al. (1998) restored separate one-hectare freshwater wetlands using each approach. They hypothesized that the two wetlands would be similar in structure and function in the long term and measured attributes of water quality, vascular plants, algae, invertebrates, and birds. The wetlands are still developing, but Mitsch et al. (1998) recommend that planting may not be needed for wetlands where plant propagules are abundant and species easily establish on their own. Large, contiguous patches of nearby natural wetlands will serve as a propagule source for colonization and facilitate the self-design of restored wetlands. In other types of wetlands where plant species have minimal dispersal or rarely establish on their own, additional plantings would be necessary in order to establish a diverse range of species (Sullivan 2001).

ANIMALS

While there has been intensive focus on establishing plants at restored wet-lands, in most cases the approach with animals has been to allow for their natural colonization; some refer to this as the "build it and they will come" approach. Since animals tend to be highly mobile, most restoration ecolo-gists have felt that a restored fauna will become established if 1) natural populations that can act as sources for the newly restored site are found nearby and 2) the habitat conditions at the newly restored site are appro-priate. In many cases, the establishment of **habitat corridors** will facilitate the dispersal of animals into restored wetlands. Multiple researchers have evaluated the colonization of animals at sites, with most work focusing on invertebrates and fish and fewer evaluations of bird or wildlife use of re-stored wetlands.

Invertebrate species that live in wetland soils or adjacent unvegetated mudflats are highly dependent on substrate conditions, in particular the texture and organic content of the substrate. As noted previously, many re-stored wetlands have coarser soils and less organic matter than natural wetlands. These substrate differences lead to substantial differences in the communities that develop in restored versus natural wetlands (Levin et al.

1996). Although the absolute abundance of organisms in the mud may be similar in restored and natural wetlands, restored wetlands tend to be dominated by different species, with substantial differences in feeding modes between restored and natural wetlands. These differences in feeding modes may lead to substantial differences in functions, such as food web support between natural and restored areas. Furthermore, the coarse soils in restored areas are likely to limit the ability of subsurface species to build tubes and burrows, further limiting some species' ability to thrive in restored wetlands. As soils develop over time, it is likely that there also will be gradual shifts in invertebrate communities.

One of the critical issues for fish use of restored wetlands is the presence or absence of creeks within the wetland. Creeks provide access for fish to move from adjacent deepwater areas up into the wetland. Multiple studies have shown that if access is available to restored wetlands via creeks, fish are likely to colonize quite rapidly since they easily move from one area to another. Minello et al. (1994) showed that the amount of edge or border area between creeks and the wetland is important in allowing fish and invertebrates to move into the wetland on high tides. In their study, shrimp densities were more than five times greater near the edge of experimental channels than in areas without channels. Fine-scale characteristics of creeks can also affect fish use of wetlands. Because of substantial differences in the creek networks in restored and natural salt marshes, fish communities in the restored salt marsh were also different (Williams and Zedler 1999). Future restoration projects should make a greater effort to include a range of creeks of appropriate shapes and sizes in order to encourage fish establishment.

Little research has been done to evaluate the effectiveness of restored wetlands in creating habitat for birds and other wildlife, although a recent special issue of *Restoration Ecology* focuses on this issue (see Morrison 2001). Not enough research has been done to make any generalizations about wildlife use, but the focus to date has been on creating the proper habitat conditions for particular wildlife species. Brown and Smith (1998) evaluated 18 restored wetlands in northern New York and found that the restoration projects increased bird habitat. However, this habitat did not provide the same function for birds as did natural wetlands. Restored sites lacked woody vegetation, and this likely reduced habitat value. VanRees-Siewert and Dinsmore (1996) measured restored wetlands of varying ages over a two-year period. The number of bird species that were breeding in the restored wetlands was significantly greater in older restored wetlands than in younger wetlands, supporting the idea that some wetland features may take time to develop.

ENDANGERED SPECIES AND UNUSUAL WETLANDS

The restoration of endangered species is especially problematic since rare species are likely to have very specific habitat requirements. There has been recent interest in developing our understanding of the restoration needs of endangered species (for example, Falk et al. 1996). Reintroduction of endangered species is a common component in recovery plans under the Endangered Species Act; however, many technical, ecological, and political questions remain for the successful reintroduction of endangered species (Falk et al. 1996). Even though there is an urgent need to focus on endangered species because of their small population size and because of the requirements of the Endangered Species Act, these restoration efforts also should focus on ecosystem functioning and sustainability in order to ensure that species are retained over the long term.

Finally, it should be noted that much of the focus of research in wetland restoration so far has been on a small number of wetland types, such as tidal wetlands or relatively simple freshwater wetlands. The restoration of some wetland types is likely to be a greater challenge, in part because of their ecological characteristics and in part because of our lack of knowledge about these ecosystems. In particular, vernal pools, peat bogs, and fens will be difficult to restore because of their unique hydrology, soil, and vegetation (National Research Council 2001).

Conclusion

It is clear that the restoration of fully functioning, sustainable wetland ecosystems is a real challenge. While we have made many advancements in wetland restoration over the past three decades, restoration remains an experimental science, with many unknowns and a high degree of unpredictability. Some of the specific improvements that are needed for the restoration of wetlands include the following:

1. *Developing clear goals and plans for restoration.* We should realize that the goals for individual projects need to be realistic and flexible (Ehrenfeld 2000). Frequently, one of the problems with high-profile restoration efforts is that our expectations for these sites are too high. An individual wetland is unlikely to provide all the ecosystem functions outlined in this chapter, and it is unrealistic to expect that a restored wetland would achieve this wide range of function. In addition, we need to realize that the restoration of wetlands takes time. As discussed in this chapter, some characteristics of wetlands (especially soils) may take many decades or even centuries to develop (Craft et al. 1999). We should not fool ourselves into thinking that functions related to these characteristics will be restored quickly. Finally, we should be flexible in our management approach, focusing on the use of adaptive management in restoration (Thom 2000).

2. *Focusing on large-scale restoration.* There is growing interest in large-scale restoration projects, such as the emphasis in watershed-based restoration projects (Williams et al. 1997). As we learn more about the connection of activities throughout watersheds and across ecosystems, it becomes clear that restoration of large blocks of habitat will be more effective in restoring ecosystem functions.

3. *Identifying limiting factors and developing better restoration techniques.* In order to develop better restoration methods, first we must improve our understanding of the limiting factors for wetland development, including both physical and biological limits. With this understanding, we will be able to improve restoration techniques. Because of the constraints of mitigation, many wetland restoration efforts focus entirely on tried-and-true methods, with little interest in experimenting with new restoration techniques, despite the fact that established methods

may not be the most effective. Greater effort should be made to develop more effective restoration methods that include integrative approaches.

4. *Developing better assessment methods.* Most assessments of restored wetlands have focused on few parameters, with data collected over the short term (Zedler and Callaway 2000). Future assessment efforts should focus on multiple parameters over the long term, as it is likely that individual parameters will differ in their evaluation of restoration progress. Long-term monitoring is needed given the time scale at which restored wetlands develop.

   In addition, the focus to date has been on assessment methods that measure ecosystem structure rather than function despite the clear understanding that the goal of restoration efforts is to improve ecosystem function. Primarily this is due to the fact that ecosystem function is much more difficult to assess than structure, and most restoration projects do not have the resources to conduct detailed evaluations of ecosystem function. More effort is needed to develop relatively simple assessment methods that reflect ecosystem function.

5. *Using a scientific approach.* Restoration projects offer a great opportunity to develop insight into ecosystem dynamics while improving restoration and management efforts at the same time. Too often in the past, we have learned from restoration efforts on a trial-and-error basis. In order to improve our understanding of wetland restoration, we need to use a scientific approach in addressing restoration challenges. Experiments should be incorporated into restoration projects, with clear experimental designs, replication of treatments, specific hypotheses, and other components of scientific experimentation. With this approach, we can more quickly improve the science of restoration ecology.

6. *Considering social aspects of restoration.* In addition to improving ecosystem functions, restoration projects can involve the public and increase their understanding of ecosystems. This is especially important in urban settings, where the need for both ecosystem restoration and

environmental education is the greatest. Efforts should be made to increase public involvement in restoration projects. In the long run, sustainability of restoration efforts depends on public support, and the public will value only a resource that they understand and appreciate.

7. *Acknowledging the challenge of restoration.* Finally, it should be acknowledged that the restoration of wetland ecosystems remains experimental. Predicting the development of restoration projects is difficult (and sometimes impossible). As we evaluate sites, we should realize that particular locations may not develop exactly as planned. In order to compensate for this uncertainty, we need to incorporate some level of adaptability into the design and management of restoration projects (Thom 2000). The challenge of wetland restoration also points to the value of existing wetlands and the need to increase efforts to conserve wetlands that already provide valuable ecosystem functions.

REFERENCES

Allen, E. B., W. W. Covington, and D. A. Falk. 1997. Developing the conceptual basis for restoration ecology. *Restoration Ecology* 5:275–76.

Alpert, P., F. T. Griggs, and D. R. Peterson. 1999. Riparian forest restoration along large rivers: Initial results from the Sacramento River Project. *Restoration Ecology* 7:360–68.

Bradshaw, A. D. 1985. Restoration: An acid test for ecology. In *Restoration ecology.* Edited by W.R. Jordan III, M. Gilpen, and J. Aber. Cambridge: Cambridge University Press, 23–29.

Brown, S. C., and B. L. Bedford. 1997. Restoration of wetland vegetation with transplanted wetland soil: An experimental study. *Wetlands* 17:424–37.

Brown, S. C., and C. R. Smith. 1998. Breeding season bird use of recently restored versus natural wetlands in New York. *Journal of Wildlife Management* 62:1480–91.

Cairns, J., Jr., ed. 1995. *Rehabilitating damaged ecosystems.* Boca Raton, Fla.: Lewis.

Callaway, J. C. 2001. Hydrology and substrate. In *Handbook for restoring tidal wetlands.* Edited by J. B. Zedler. Boca Raton, Fla.: CRC Press, 89–117.

Casagrande, D. G., ed. 1997. *Restoration of an urban salt marsh: An interdisciplinary approach.* New Haven, Conn.: Yale University Press.

Chimney, M. J., and G. Goforth. 2001. Environmental impacts to the Everglades ecosystem: A historical perspective and restoration strategies. *Water Science and Technology* 44:93–100.

Craft, C., J. Reader, J. N. Sacco, and S. W. Broome. 1999. Twenty-five years of ecosystem development of constructed *Spartina alterniflora* (Loisel) marshes. *Ecological Applications* 9:1405–19.

Dahl, T. E. 1990. *Wetlands losses in the United States, 1780's to 1980's.* Washington, D.C.: U.S. Department of the Interior, Fish and Wildlife Service.

Dobson, A. P., A. D. Bradshaw, and A. J. M. Baker. 1997. Hopes for the future: Restoration ecology and conservation biology. *Science* 277:515–22.

Ehrenfeld, J. G. 2000. Defining the limits of restoration: The need for realistic goals. *Restoration Ecology* 8:2–9.

Falk, D. A., C. I. Millar, and M. Olwell, eds. 1996. *Restoring diversity: Strategies for reintroduction of endangered plants.* Washington, D.C.: Island Press.

Galatowitsch, S. M., and A. van der Valk. 1994. *Restoring prairie wetlands: An ecological approach.* Ames: Iowa State University Press.

Hammer, D. A. 1997. *Creating freshwater wetlands.* Boca Raton, Fla.: CRC Press.

Jenny, H. 1941. *Factors of soil formation: A system of quantitative pedology.* New York: McGraw-Hill.

Kadlec, R. H., and R. L. Knight. 1996. *Treatment wetlands.* Boca Raton, Fla.: Lewis.

Kentula, M. E., R. P. Brooks, S. E. Gwin, C. C. Holland, A. D. Sherman, and J. C. Sifneos. 1992. An approach to improving decision making in wetland restoration and creation. EPA/600/rd-92/150. U.S. Environmental Protection Agency, Corvallis, Oregon.

Kusler, J. A., and M. E. Kentula, eds. 1990. *Wetland creation and restoration: The status of the science.* Washington, D.C.: Island Press.

Langis, R., M. Zalejko, and J. B. Zedler. 1991. Nitrogen assessment in a constructed and a natural salt marsh of San Diego Bay. *Ecological Applications* 1:40–51.

Leopold, A. 1966. *A Sand County almanac. With other essays on conservation from Round River.* New York: Oxford University Press.

Levin, L. A., D. Talley, and G. Thayer. 1996. Succession of macrobenthos in a created salt marsh. *Marine Ecology Progress Series* 141:67–82.

Lindau, C. W., and L. R. Hossner. 1981. Substrate characterization of an experimental marsh and three natural marshes. *Soil Science Society of America Journal* 45:1171–76.

Middleton, B. 1999. *Wetland restoration, flood pulsing, and disturbance dynamics.* New York: Wiley.

Minello, T. J., R. J. Zimmerman, and R. Medina. 1994. The importance of edge for natant macrofauna in a created salt marsh. *Wetlands* 14:184–98.

Mitsch, W. J., and J. G. Gosselink. 2000. *Wetlands.* New York: Wiley.

Mitsch, W. J., X. Y. Wu, R. W. Nairn, P. E. Weihe, N. M. Wang, R. Deal, and C. E. Boucher. 1998. Creating and restoring wetlands: A whole-ecosystem experiment in self-design. *Bioscience* 48:1019–30.

Morrison, M. L. 2001. Introduction: Concepts of wildlife and wildlife habitat for ecological restoration. *Restoration Ecology* 9:251–52.

National Research Council. Committee on Characterization of Wetlands. 1995. *Wetlands: Characteristics and boundaries.* Washington, D.C.: National Academy Press.

National Research Council. Committee on Mitigating Wetland Losses. 2001. *Compensating for wetland losses under the Clean Water Act.* Washington, D.C.: National Academy Press.

National Research Council. Committee on Restoration of Aquatic Ecosystems. 1992. *Restoration of aquatic ecosystems: Science, technology, and public policy.* Washington, D.C.: National Academy Press.

Sanville, W., and W. J. Mitsch. 1994. Preface to special issue: Creating freshwater marshes in a riparian landscape: Research at the Des Plaines River Wetlands Demonstration Project. *Ecological Engineering* 3:315–17.

Simenstad, C. A., and R. M. Thom. 1996. Functional equivalency trajectories of the restored Gog-Le-Hi-Te estuarine wetland. *Ecological Applications* 6:38–56.

Steyer, G. D., and D. W. Llewellyn. 2000. Coastal Wetlands Planning, Protection, and Restoration Act: A programmatic application of adaptive management. *Ecological Engineering* 15:385–95.

Sullivan, G. 2001. Establishing vegetation in restored and created coastal wetlands. In *Handbook for restoring tidal wetlands.* Edited by J. B. Zedler. Boca Raton, Fla.: CRC Press, 119–55.

Thom, R. M. 2000. Adaptive management of coastal ecosystem restoration projects. *Ecological Engineering* 15:365–72.

Tilman, D. 2000. Causes, consequences and ethics of biodiversity. *Nature* 405:208–11.

Toth, L. A., S. L. Melvin, D. A. Arrington, and J. Chamberlain. 1998. Hydrologic manipulations of the channelized Kissimmee River: Implications for restoration. *Bioscience* 48:757–64.

VanRees-Siewert, K. L., and J. J. Dinsmore. 1996. Influence of wetland age on bird use of restored wetlands in Iowa. *Wetlands* 16:577–82.

Weinstein, M. P., J. M. Teal, J. H. Balletto, and K. A. Strait. 2001. Restoration principles emerging from one of the world's largest tidal marsh restoration projects. *Wetlands Ecology and Management* 9:387–407.

Williams, G. D., and J. B. Zedler. 1999. Fish assemblage composition in constructed and natural tidal marshes of San Diego Bay: Relative influence of channel morphology and restoration history. *Estuaries* 22:702–16.

Williams, J. E., C. A. Wood, and M. P. Dombeck, eds. 1997. *Watershed restoration: Principles and practices.* Bethesda, Md.: American Fisheries Society.

Zedler, J. B. 2000. Progress in wetland restoration ecology. *Trends in Ecology and Evolution* 15:402–7.

———, ed. 2001. *Handbook for restoring tidal wetlands.* Boca Raton, Fla.: CRC Press.

Zedler, J. B., and J. C. Callaway. 1999. Tracking wetland restoration: Do mitigation sites follow desired trajectories? *Restoration Ecology* 7:69–73.

———. 2000. Evaluating the progress of engineered tidal wetlands. *Ecological Engineering* 15:211–25.

Zedler, J. B., J. C. Callaway, and G. Sullivan. 2001. Declining biodiversity: Why species matter and how their functions might be restored in Californian tidal marshes. *Bioscience* 51:1005–17.

SUGGESTED READINGS

Clewell, A., and J. P. Rieger. 1997. What practitioners need from restoration ecologists. *Restoration Ecology* 5:350–54.

Ehrenfeld, J. G., and L. A. Toth. 1997. Restoration ecology and the ecosystem perspective. *Restoration Ecology* 5:307–17.

Jordan, W. R., III, M. Gilpen, and J. Aber, eds. 1985. *Restoration ecology.* Cambridge: Cambridge University Press.

Mills, S. 1995. *In service of the wild: Restoring and reinhabiting damaged land.* Boston: Beacon.

Mitsch, W. J., and R. F. Wilson. 1996. Improving the success of wetland creation and restoration with know-how, time, and self-design. *Ecological Applications* 6:77–83.

Montalvo, A. M., S. L. Williams, K. J. Rice, S. L. Buchmann, C. Cory, S. N. Handel, G. P. Nabhan, R. Primack, and R. H. Robichaux. 1997. Restoration biology: A population biology perspective. *Restoration Ecology* 5:277–90.

Urbanska, K. M., N. R. Webb, and P. J. Edwards, eds. 1997. *Restoration ecology and sustainable development.* Cambridge: Cambridge University Press.

Whelan, C. J., and M. L. Bowles, eds. 1994. *Restoration of endangered species: Conceptual issues, planning, and implementation.* Cambridge: Cambridge University Press.

From Past to Present

A Historical
Perspective
on Wetlands

William B. Meyer

The background provided by history is always illuminating, but the perspective of **environmental history** has more than mere background to offer in a volume such as this. It has things to say about today and tomorrow as well as about yesterday. Above all, we can learn from history how much of what seems obvious at any one time did not—and perhaps will not—always seem so under the different circumstances of other times. What we take for granted earlier Americans did not. That could, of course, give us the chance to look down on them as ignorant or shortsighted or stupid. We can also discover, however, that they were not simply wrong in thinking what they did. By the same token, the future may look at us—its own past—with as much puzzlement as we feel at first when we look at much of our past. Each is likely to ask, How could those people possibly have thought and acted as they did? Trying to answer that question about others is a good way to understand how it might apply to us as well. We have good reasons for our beliefs, but so did people who thought differently from us.

Everything that is taken for granted, and necessarily so, in the chapters included in this volume represents the point of view of one place and period. Much of it may one day seem outdated and wrongheaded, just as it now makes much of the past seem so. The future may well look at wetlands, the services that they provide, and the means of regulating their use and abuse in ways other than those of today's received wisdom. For the past certainly viewed them in other ways, and history and change have not come to an end with us.

Environmental History: Concepts and Approaches

Environment, though the word has many meanings, stands here for the biophysical surroundings of human life. History is the study of how and why human life and human societies have differed from one time to another and have changed over time. Environmental history is the study of how the relations of human life and human societies to their biophysical surroundings have changed throughout the past and up to the present. Those relations include both human impacts on the environment and the ways in which the environment affects or constrains human actions.

What sets environmental history (and history more generally) apart from the other approaches represented in this volume is not simply the fact that it deals with the past and the others deal with the present and the future. It takes an integrative approach, bringing together issues of ecology, economics, politics, and culture that elsewhere tend to be treated separately. It pays attention to the ways in which all these factors affect one another. It tends to avoid abstract generalizations or the search for universal laws that are supposed to hold true for human behavior in all times and places. Instead, it pays particular attention to the specificities of time and place and to the ways in which human actions and their underlying motives differ with locale and period. What it tries above all to avoid is what historians call **presentism** or anachronism, or the inappropriate use of present-day concerns or beliefs to understand or sometimes simply to condemn or to praise past societies. The particular danger that environmental history faces in studying the past is that of becoming a simplistic environmentalist history. It can easily be warped or slanted by the concerns of modern environmental activism—with which most environmental historians are deeply sympathetic—into praising or blaming people of the past simply for the degree to which they seem to have conformed to it.

Environmental history, then, tries to understand the environment's relation to human life and livelihood in particular settings. In doing so, there is no better way to start than to look first at the environment's practical importance to its occupants. In these terms, any feature of the environment can be either beneficial or detrimental; it can offer benefits or impose costs. It can represent a help or a hindrance, a resource or a hazard, or it can represent neither and be disregarded altogether. A resource is anything that can help us do things that we want to do. Petroleum is a natural resource that is so vital to everyday transportation in modern society that any halt in its extraction and refining would seriously disrupt life as we know it. A hazard is anything that threatens things that we value or that gets in the way of things that we want to do. Snow is a serious hazard to transportation—particularly to driving and to flying—just as petroleum is an essential resource for it. Snow is also a resource for skiing and other forms of recreation, just as tanker accidents and spills make oil a hazard for ocean fisheries.

The term **environmental services** has recently been introduced to describe better than the narrower-sounding one of "resources" the whole range of benefits that features of the environment may provide (Daily 1997). They include many that help us even though we are not much aware of them or do not consciously exploit them in the way that we extract, refine, and consume a resource such as petroleum. Important services provided by the environment that we do not usually classify as natural resources include the regulation and purification of water flow, habitat for flora and fauna, and regulation of air quality and microclimate. So too, the parallel term of **environmental disservices** is a useful one for the many ways in which features of the environment impose some expense or hardship or inconvenience even if they are not obvious hazards on the order of storms or earthquakes. Any feature of the environment, especially the more common ones, can and most likely will provide many services and many disservices at the same time. Though practical considerations usually come first, they are not the only dimension of human–environment relations. The services and disservices that the environment provides also include spiritual and aesthetic value and any spiritual or aesthetic offense that they give.

Why is it that any feature of the environment at any time is a resource or a hazard, a source of services and/or of disservices? Not because of its own qualities alone, not because it is inherently good or bad, but because of the manner in which its own qualities interact with the human ways of life with which it comes into contact. The demands of the latter are what make aspects of the environment useful or harmful. As a result, usefulness and harmfulness are not eternal and fixed characteristics but ones that can change greatly over time. Petroleum was useless and valueless until the invention first of kerosene lighting and then of the internal combustion engine made it an indispensable resource. It did not become valuable because it had undergone any change itself. It had the same chemical composition and properties as before, but their significance was different. It may again become valueless, or at least much less valuable than it is now, long before we exhaust its supplies if cheaper and cleaner sources of energy are devised and developed. Snow was not always a hazard or a hindrance to transportation in the northern United States but was once of great help to it. Before the coming of motorized vehicles and paved roads, a snow-covered surface could be traversed by sleigh or sled much easier than the muddy, rutted dirt roads of spring, summer, and fall. Winter was the season of the speediest travel and the easiest and cheapest movement of freight over land (Meyer 2000).

It follows that anything in the environment can become or can cease to be a help or a hindrance simply because the human ways of life around it have changed. It can also, of course, do so because it has been altered itself. But what matters in the environment in one period of history, even if it remains physically unchanged, will not always matter or matter in the same way in another period. American geographers in the early 20th century coined the term **sequent occupance** to describe this phenomenon. This is "the principle that the significance of the physical environment changes as attitudes, objectives, and technical abilities of the people living in it change" (White and Foscue 1954, 6–7). A single period of occupance of an area is one in which the roles played by any feature in it are more or less constant. A new stage in the sequence begins when they change for one reason or another.

Historians and geographers have written studies of the sequent occupance of many environments and of how their use and evaluation have changed as ways of life have changed. Some of the best of them deal with wetland environments in the United States. These detailed local histories show many sequences of change. But there is one sequence that is common to almost all of them. It forms the basic pattern of American wetland history. In aboriginal times, before European settlement, wetlands, on the whole, were valued. Among the white settlers, on the whole, and well into the 20th century, the dominant opinion strongly disapproved of wetlands and supposed that the best thing that could be done with them was to drain them, destroying their special character to convert them into dry land. No less widespread today is the belief that the best thing that can be done with wetlands is to preserve them in as nearly their unaltered state as possible and even to create new ones to replace those that have been lost. In order to understand why these changes occurred, the first question that the environmental historian asks is, What changes have taken place in the American way of life that may have tended to bring them about?

Wetland History

The first human beings to live in and among the wetlands of the present-day United States were the aboriginal occupants of the land. Different peoples used wetlands in many different ways, but most found them, on the whole, much more of an asset than a liability. They were not, for one thing, a serious hazard to health. They did not, in pre-Columbian times, harbor the debilitating mosquito-borne disease of malaria. Scholars generally agree that the microscopic parasite that causes malaria was not present in the Americas until European explorers and settlers inadvertently brought it over from the Old World. Nor was yellow fever, likewise carried by mosquitoes and transmitted by their bite. For peoples in whose livelihood hunting played a major part, wetlands were highly valuable for the wealth of game—fish, shellfish, migratory birds, and fur-bearing animals such as beaver and muskrats—that they harbored. Useful plants, such as wild rice in some regions, were also collected. Rich wetland soils were cultivated for domesticated crops.

If Amerindians on the whole maintained wetlands as they were, it was not out of some mystical reverence for nature but rather because wetlands, in more or less their natural state, suited their way of life. They were quite ready to alter them when it proved otherwise. Extensive remnants of ridge-and-furrow fields among wetlands in the upper Midwest offer a case in point. They were raised and converted to dry ground so that corn and other crops could be better grown than in the saturated soils around them and to give corn—near the northern limits of its cultivability—added protection from ground frosts (Gallagher et al. 1985).

Many tribes, finally, sought wetlands as a defensible refuge from attacks or reprisals during the period of European conquest and settlement. The last battles of the war that they waged against the colonists of southern New England in the 1670s were fought in the swamps of Rhode Island. Those of the Black Hawk War of the 1830s took place in the wetlands of the upper Midwest. The Seminole Indians, whose modern way of life is closely tied to the landscape of the Everglades and the adjoining Big

Cypress Swamp of southern Florida, were latecomers to the area. They migrated from the north in the early 19th century in search of a stronghold where they could resist the encroachments of white settlers.

None of this made the Euro-American inhabitants of North America like wetlands any better, but they already had many other reasons for disliking them. Their attitudes toward them, on the whole, were negative, indeed quite strongly so. For a number of reasons, including both environmental change and differences in ways of life, wetlands were much less unambiguously an asset to the ways of life of the new settlers than they had been to the aborigines.

The large majority of white Americans farmed for a living, but few of them practiced methods suitable to wetlands. As a result, they looked on marshes and swamps as little more than useless wastelands. Only a few activities, such as collecting hay for livestock fodder in coastal marshes and growing rice in the flooded fields of the southern Atlantic and Gulf coasts, took advantage of the opportunities that they offered. Even the growing of rice developed less from Anglo-American initiative than from the skills of slaves who brought their expertise in wetland cultivation from the coastal swamps of West Africa (Carney 1993).

Wetlands also posed serious obstacles to peacetime travel, especially by horse, carriage, and wagon—problems unknown to the Native Americans before Columbus, who had not possessed or used such vehicles. In the words of the historical geographer Hugh Prince, "Natives were able to paddle small canoes along narrow channels through reeds and rushes, or leap fleet-footed from tussock to tussock. White travelers and their horses had to wallow in mud and water" (Prince 1997, 118–19). With modern paving methods still unavailable, good roads were almost impossible to construct through marshes and swamps.

Not only in the countryside but around cities too, the transplantation of European ways made wetlands much more of a hindrance than a help. Aboriginal settlements had moved seasonally or at intervals of a few years; flooding in low-lying river bottomland was much more of a hazard for the permanent towns and cities of the newcomers. Along urban shorelines, wetlands blocked the approach of shipping to the shore yet were useless for construction. Much of the nearby lowland that growing cities needed for their expansion was inhospitable wetland. The population of early 19th-century Boston crowded into the narrow peninsulas, while large stretches of marsh and tidal flat, offensive to the eye and nose at low water, extended far out to sea. Such cities as Chicago and New Orleans could only grow through massive reclamation and landfill. On top of their other failings, wetlands also seemed ugly, repulsive, and forbidding, the haunt of reptiles and vermin. Most Americans' tastes in landscape had no place for such scenery. There were a few who found them beautiful. Henry David Thoreau, the landscape painter Martin Johnson Heade, the naturalist John Muir, and the southern poet Sidney Lanier are among the best known (Miller 1989). But disgust and aversion were the usual reactions to wetlands. They were rarely even called wetlands (though sometimes "wet lands") until the 20th century but rather swamps, morasses, mud holes, and other terms that were often used figuratively as ones of disparagement and symbols of evil.

But far and away the most important reason for the widespread dislike and fear of wetlands in 17th-, 18th-, and 19th-century North America was their association with disease. Once it was introduced, malaria spread and established itself so quickly across an enormous expanse of the continent that the newcomers had little reason to suspect that it had not always been present. It was known by many other names, such as intermittent fever, fever and chills, and the ague, but by any name it was a feared, debilitating, and most unpleasant illness. The name malaria, meaning "bad air," captured the widespread belief as to its cause. It was generally blamed on the effects of miasma, a gaseous substance supposedly given off from stagnant waters and decaying vegetation. Though mistaken, the belief was far from an unreasonable one. For in fact malaria was most prevalent in and among wetlands, which should logically have been leading sources of miasma and where the mosquitoes that actually transmitted it had their prime habitat. It was also most prevalent in the South with its warm climate, where decay and the generation of miasma would logically have been most intense.

The medical fears of wetlands were especially acute in the South and in the Midwest, whose vast reaches of wet prairie made malaria almost impossible to escape over large regions. Yet it was a problem of national proportions. It was found as far north as upstate New York and southern New England into the early 20th century. Much of the hinterland of 19th-century New York City—the fringes of the New Jersey Meadowlands, most of Staten Island, and parts of the Hudson valley—could have been occupied as suburbia long before it was except for its well-deserved reputation for malariousness.

The obvious thing to do was to turn wetlands into solid ground. Coastal flats could be raised by landfill. Channels could be dug through inland swamps and marshes to carry off their excess water into local streams and rivers. And it was a doubly attractive measure that seemed likely both to improve the health of the surrounding areas and to pay for itself or more by making useful land for either farming or construction out of a useless waste. Given what was believed about its origins, it seemed that miasma could not be eradicated by any means less drastic. But drainage would also improve the look of the area, by the tastes of the time, and by bettering the land and reducing sickness, it would do much to secure the interests and well-being of future generations. The government reclamation expert Marshall O. Leighton summed up the case for drainage with a striking analogy in the early 20th century. He likened America's wetlands to "a wondrously fertile country inhabited by a pestilent and marauding people who every year invaded our shores and killed and carried away thousands of our citizens, and each time shook their fists beneath our noses and cheerfully promised to come again" (Palmer 1915, 1). What response was possible, he asked, other than to conquer and subdue such an enemy?

To make much headway, however, the drainage movement had to make advances in both technology and law. The first came through such steps as the development and manufacture of cheap drain tile and of mechanized earthmoving equipment. The legal problems that had to be overcome included perhaps the chief obstacle to drainage on any but the most local scale. The area covered by most wetlands of substantial size, plus the adjoining areas through which drainage channels might have to run, was generally fragmented into parcels belonging to many different owners. Yet the wetland, as a single entity, could be drained only by a single coordinated project cutting across parcel boundaries. No one part could be drained in isolation, for it would simply be reflooded by waters from the others. But private enterprise could not make the needed arrangements. The costs and difficulties in getting all the owners involved to agree by negotiation on the details of the work and the allotment of the costs were generally severe enough to keep anything from being done. The American forest could be cleared and converted to farmland, tree by tree, by separate pioneer families living and working alone. The prized national principles of individualism and private effort broke down at the marsh's edge. Wetlands of any significant size, it became clear, could not be converted without aggressive government intervention—just as we assume today that they cannot be preserved without it. Those landowners reluctant to take part had to be coerced into doing so, but it aroused few qualms because of the general assumption that it was for their own as well as for everyone else's good.

Thus, state after state began to pass ditch laws, which guaranteed would-be drainers the right to dig their channels across the private lands of others when necessary and to be reimbursed for any good that they did adjoining lands. Later came drainage district acts. These laws formed local government bodies that could undertake large reclamation projects and assess all the landowners who benefited from their share of the costs through taxation. When the Ohio legislature enacted such a law in 1853, a newspaper hailed it with the words, "Now, citizen farmers, you have a law that will oblige all to pay their share of the expense, and if you must first drain other lands before you can drain yours, the owners must pay for it" (*Perrysburg [Ohio] Journal* 1853, 2). The law was challenged in 1860 before the state supreme court. The judges upheld it, reasoning that "the execution of these works is beyond the power of isolated individual effort, and that the public authority must be invoked . . . to override the conflicts of individual opinion and selfishness" (*Reeves v. Treasurer of Wood County* 1859).

Such laws still had one more obstacle to pass. Many authorities still insisted that private rights in land could be invaded only for a distinctively "public purpose." Bettering the public health was plainly one such purpose, and laws that stated that aim generally passed muster. It was less clear that simply improving land for agriculture for private owners, however many of them might benefit, passed the same test. Over time, however, the state and federal courts began to agree that it did and that the constitutionality of a ditch or drainage district act did not depend on proof of its benefits to health. Thereby, as one expert wrote in 1915, "the beginnings have been laid for a rule to the effect that one should not refuse or neglect to use or improve his own land, where such non-use will prevent others from using or improving their land or property" (Palmer 1915, 39).

Given the tools that they needed and convinced that the benefits of drainage would repay its costs many times over, Americans threw themselves into the work. Such projects as the state of Florida's persistent efforts to drain the Everglades dwarfed any previous efforts at government transformation of the environment. Drainage advanced along with the irrigation of arid lands, with which it had many similarities. Like drainage, irrigation required cooperative and coordinated action: the one to supply water from a common source, the other to remove it. They were commonly lumped together as "land reclamation," and the same economic, legal, and political developments worked to the benefit of both. By the 1920s, a vast area of the United States lay within the boundaries of drainage and irrigation districts.

The era of drainage gave way in the course of the 20th century to the era of wetland preservation and creation in which we now live. Much of it is dealt with in the following chapters of this volume, but this history would be incomplete without some discussion of how it came about. It occurred as the combined result of many developments.

American agriculture, which had enjoyed boom times through World War I, entered a severe crisis of low prices and overproduction that lasted throughout the 1920s and 1930s. Drainage had once seemed almost certain to turn a profit. Now the farmers in many drainage districts found themselves burdened with the project costs and at a serious disadvantage in competing with others. Many districts went bankrupt during the Great Depression. Much of the land that they had drained was abandoned. Much of it had in any case turned out to be poorer than promoters had expected. The peat lands of the northern Midwest, for example, once dried out, had proved poor soil for farming and a perpetual fire hazard during drought years. In 1948, the ecologist Aldo Leopold wrote a memorable essay, "Marshland Elegy," collected in his *Sand County Almanac* that vividly contrasted the health and productivity of the region's wetlands with their impoverished and worthless state after they had been drained.

As agriculture increasingly became mechanized, the drastic 20th-century decline in the total and proportional American population that lived by farming also meant the emergence of new perspectives on proper land use. Among the rising numbers of amateur hunters, sportsmen, and naturalists, concern grew in the early 20th century about the decline in numbers of migratory waterfowl and other species as a result of the loss of their wetland habitat to drainage projects. The value of anything is likely to rise as it becomes more scarce, and the populations of wild game that had once been too plentiful to worry about depleting no longer seemed as inexhaustible as they once had. During the 1920s and 1930s, the syndicated cartoonist Jay N. Darling did much to awaken Americans to the threat that drainage posed to the habitat of migratory game birds. Serving briefly in the Roosevelt administration at the start of the New Deal, he helped create the federal Duck Stamp program, through which hunters and conservationists provided money to form, maintain, and expand a system of national wildlife refuges, much of it wetland salvaged from marginal farm operations.

No less important was the discovery at the beginning of the 20th century that malaria was caused not by miasma but by mosquito-borne pathogens. This was a discovery of immeasurable importance for wetland management. It discredited the miasma theory, according to which the very existence of marshes and swamps almost inevitably meant illness for their neighbors. It meant that the disease might be eradicated without any need for drainage by measures aimed at the insects themselves. Some of these measures, of course, could do serious damage themselves. The post–World War II use of the insecticide DDT was phenomenally successful in mosquito control in the United States and abroad. Rachel Carson's 1962 book *Silent Spring*, calling attention to its harmful effects on songbirds and other species, remains a landmark in the rise of environmental awareness.

The spread of popular environmentalism fostered by Carson and others has instilled in many Americans the conviction that land is always better in its natural state than it is if altered in any substantial way by human action. Drainage, from once being assumed to be an improvement, is now assumed to be damage unless proven otherwise. Regionalists such as Florida's Marjory Stoneman Douglas helped through their writings to change the public's image of wetlands—in her case, the Everglades—from ugly morasses to landscapes of beauty. Ecologists such as Eugene Odum of the University of Georgia have demonstrated that wetlands, particularly coastal ones, are not the useless wastelands of long-standing legend but, on the contrary, unusually rich and productive ecosystems on which many other elements of the environment depend.

Other services—new or newly recognized—that wetlands provide have been revalued upward. As water pollution has become a growing problem, the purification of stream flow by wetlands has become a newly prized asset. So has their contribution to a stable water supply, as southern Florida belatedly discovered when success in drainage made drought crises more frequent and more severe than before. Their storage capacity in times of high flow now matters more too as the dollar cost of floods continues to increase. The postwar decades have seen steadily denser and more valuable development of exposed lowlands along rivers. That wetland drainage could increase flood heights downstream was widely understood even in the 19th century. It was simply accepted as a necessary cost of the operation. Now the mounting toll of damage and of federal flood disaster relief has caused Washington to pay new attention to ways of reducing both by preserving wetlands.

Nor is it likely that the shift in perceptions and in actions would have been so complete if wetland preservation did not offer many Americans additional attractions. The environmentalist case for wetland preservation is a sound one and one to which the large majority of the public seems to have been converted. Yet it is no more sound than many other environmentalist claims and warnings to which the public and the legislatures continue to turn a deaf ear. Global warming presents serious threats and challenges, but the most effective measures against it—such as carbon taxes—would be a financial burden on all Americans who drive—that is, on almost all Americans. It is probably no coincidence that no such measures have been implemented in the United States. The impact of wetland preservation regulations, by contrast, is highly concentrated. It falls on the relatively few who own wetlands that could be developed. The benefits are more widely dispersed; they accrue in particular to towns and home owners who, in an era when landscape amenities are menaced by sprawl and development, wish to maintain open space and fend off unwanted construction in their environs by means that cost them nothing.

Yet the pressure for development continues, and to date it has more than offset the effects of changed attitudes. Not only has population grown, but a number of additional factors have magnified its impact on land use in the postwar United States. Households have grown smaller and thus more numerous, second houses have multiplied, and large house lots have become more common. Further, the very rise in environmental appreciation means that areas rich in attractive landscapes—which now includes wetlands—are likely to feel the greatest pressure of development.

The net human impact on America's wetlands has mostly been loss. The figure generally accepted is that somewhat more than half the wetlands originally present in the United States have been destroyed. They covered some 345,000 square miles, about 11% of the lower-48-state area in the 18th century, before impact had been significant. They have now been reduced to about 5% of the same land area (Schmid 2000). Large, individual wetland areas, such as Ohio's Black Swamp, have disappeared altogether. There has been some modest degree of restoration in recent years while failing thus far to keep pace with additional destruction. Policies favoring preservation have clearly slowed wetland loss, though they have not stopped it. That they have measurably slowed it, however, is itself noteworthy given that it is now physically easier than ever to convert wetlands to dry. It took the drainage movement some time to translate its ideals into action; the same is true of the preservation movement.

Conclusion

Why have American interactions with wetlands changed so dramatically over time? The easiest answer—so easy that it needs to be looked at with some suspicion—is that today's concern with preservation is simply part of the rise of environmental consciousness. Modern environmentalism is a political and philosophical stance that involves a number of beliefs. One of them is that the biophysical environment matters in human life and that maintaining or improving its quality is therefore important. Another is that such considerations as human health and environmental aesthetics need to be considered as well as more narrowly practical matters. A third is that we should consider the interests and the welfare of future generations in our decisions as well as the immediate needs of the present. Still another is that private actions may need to be regulated by the community to ensure better outcomes.

Today, these beliefs, when taken together, point toward preserving wetlands. What is interesting is that the same beliefs at the beginning of the 20th century, set in the context of the time, all pointed toward draining wetlands. The drainage crusaders, in these terms, were among the leading American environmentalists of their times. They were not stupid or greedy. They believed that drainage meant environmental improvement, including better human health and a more attractive and healthy ecosystem. They believed too that they were acting in the future's best interest at least as much as in their own, bequeathing a better landscape to their children and grandchildren. They believed that those who resisted drainage were acting selfishly and antisocially and that the rights of private property must to some degree give way to the higher common good.

How could the same principles now point to such a radically different course of action? It is because the circumstances within which they are applied have changed. The history of American wetlands is that of their shift from service to disservice to service as American life has changed. Their future, of course, is an open question.

REFERENCES

Carney, J. A. 1993. From hands to tutors: African expertise in the South Carolina rice economy. *Agricultural History* 67(3):1–30.

Daily, G. C. 1997. *Nature's services: Societal dependence on natural ecosystems.* Washington, D.C.: Island Press.

Gallagher, J. P., R. F. Boszhardt, R. F. Sasso, and K. Stevenson. 1985. Oneota ridged field agriculture in Southwestern Wisconsin. *American Antiquity* 50:605–12.

Meyer, W. B. 2000. *Americans and their weather: A history.* New York: Oxford University Press.

Miller, D. C. 1989. *Dark Eden: The swamp in nineteenth-century American culture.* New York: Cambridge University Press.

Palmer, B. 1915. *Swamp land drainage with special reference to Minnesota.* University of Minnesota Studies in the Social Sciences 5. Minneapolis: University of Minnesota.

Perrysburg (Ohio) Journal. 1853. May 16.

Prince, H. 1997. *Wetlands of the American Midwest: A historical geography of changing attitudes.* Chicago: University of Chicago Press.

Reeves v. Treasurer of Wood County. 1859. 8 Ohio S.Ct. 333, 344.

Schmid, J. A. 2000. Wetlands as conserved landscapes in the United States. In *Cultural encounters with the environment: Enduring and evolving geographic themes.* Edited by A. B. Murphy and D. L. Johnson. Lanham, Md.: Rowman & Littlefield, 133–55.

White, C. L., and E. Foscue. 1954. *Regional geography of Anglo-America.* 2nd ed. Englewood Cliffs, N.J.: Prentice Hall.

SUGGESTED READINGS

Blake, N. M. 1980. *Water into land: A history of water management in Florida.* Tallahassee: University Press of Florida.

Kirby, J. T. 1995. *Poquosin: A study of rural landscape and society.* Chapel Hill: University of North Carolina Press.

Outwater, Alice. 1996. *Water: A natural history.* New York: Basic.

Prince, H. 1997. *Wetlands of the American Midwest: A historical geography of changing attitudes.* Chicago: University of Chicago Press.

Sullivan, R. 1998. *The Meadowlands.* New York: Scribner.

Vileisis, A. 1997. *Discovering the unknown landscape: A history of America's wetlands.* Washington, D.C.: Island Press.

Muddy Policies and Tidal Politics

THE POLITICS
OF WETLANDS

Mary A. Hague

In central Pennsylvania, the sounds of banjo frogs and red-winged black birds resonate over the cattails of a wetland, the flowers of moth mullein and teasel add muted color, and a little green heron pecks gracefully for its next meal. The wetland provides not only habitat for the little green heron but also flood protection and water purification. The complexity of the politics of wetlands protection matches the richness of the wetland's ecology and therefore can be studied as an ecology in its own right. Interacting institutions, interdependent actors, and "environmental" factors, such as the **political culture** and **partisan** climate, shape this political ecology and influence wetlands policy. This chapter explores how the political ecology responds to conflicting pressures regarding

101

wetlands policy, providing a brief policy history and introducing one approach to policy study.

Brief Overview of Wetlands Regulation

Public policy includes the laws, rules, and regulations made and implemented by the legislative, executive, and judicial branches of local, state, and national governments. Bureaucracies at all three levels of government implement these policies. Environmental policies are public policies designed to secure the conservation, preservation, and remediation of natural resources and in many cases the protection of public health. Several national environmental polices protect the wetlands of the United States by regulating activities such as dredging and filling of wetlands. Among such policies are the **River and Harbors Act of 1899**, various **Farm Bills** with "Swampbuster" provisions (which tie agricultural subsidies to wetlands conservation), and the **North American Wetlands Conservation Act**. These policies rely on both regulatory and nonregulatory tools, such as tax credits and subsidies, to protect and conserve wetlands. The most significant policy addressing wetlands, however, and the subject of this chapter is the **Clean Water Act** (CWA) of 1972 and its amendments.

The CWA regulates the nation's surface waters by establishing ambitious water quality goals and creating a permitting system to control discharges into U.S. waters. Section 404 of the CWA authorizes the **Environmental Protection Agency** (EPA) to regulate the discharge of dredged or fill material into navigable waters. Through administrative, legislative, and judicial decisions, navigable waters have come to include wetlands. "For regulatory purposes under the Clean Water Act, the term wetlands means 'those areas that are inundated or saturated by surface or ground water at a frequency and duration sufficient to support, and that under normal circumstances do support, a prevalence of vegetation typically adapted for life in saturated soil conditions. Generally include swamps, marshes, bogs and similar areas'" (U.S. Environmental Protection Agency 2001). Much of the politics of wetlands regulation revolves around fine-tuning this general regulatory definition through wetlands delineation.

The permitting program created to implement Section 404 is administered by the **U.S. Army Corps of Engineers** (hereinafter Corps); the EPA, however, retains oversight through review and comment and has the power to veto specific permits. Therefore, plans to drain and fill a swamp, bog, or prairie pothole on one's property may require a Section 404 permit issued by the Corps. The permitting process requires that applicants submit information about the environmental impact on wetlands of the proposed activity and demonstrate that there are no practicable alternatives that avoid such impacts. While Section 404 applies to private and public property, most of the political debate over Section 404 has resulted from its application to private property because denials of these permits prevent development that might otherwise increase the economic value of the property.

The CWA's Section 404 as implemented by the Corps and the EPA faces several criticisms. Environmentalists argue that the permitting criteria are too lenient, allowing for the destruction of too many wetlands. They also claim that the existing regulations, weak as they are, are insufficiently enforced and that permits are often approved without sufficient regard to environmental damage to wetlands (Gaddie and Regens 2000). On the other hand, private property activists claim the opposite: that the regulations are far too strict; that enforcement efforts are unnecessarily heavy-handed; that the regulations change often, creating confusion; and, most important, that the regulations imposed "take" private property through the reduction of property value when development or farming is prohibited (Gaddie and Regens 2000; Miniter 1991). Additionally, in the area of wetlands protection, it is often difficult for property owners and their lawyers to anticipate how the courts will interpret wetlands regulations. These criticisms generated a wave of legislative and political activism during the 1990s, as will be seen in this chapter.

Box 5.1
The Individual Permit Process

1. Apply to district or division office of the U.S. Army Corps of Engineers
   A) Pre-application consultations for major projects;
   B) Assemble application, including a detailed description of proposed activity, accompanied by cross-section, bird's-eye, and vicinity maps
2. Public notice and comment period
   A) The Corps identifies potentially interested parties and appropriate means of public notice, usually government offices.
   B) During this time period, the Corps and other state and local agencies, interest groups and interested individuals, review the application.
   C) Public hearings may be held as a result of substantial comments or citizen requests.
   D) Lack of response from the public understood as approval and absence of objection.
3. Permit Evaluation by the Corps
   Permit evaluation considers a number of characteristics of the proposed activity, including:
   a. current regulatory prohibitions on discharges;
   b. application of any existing general permit;
   c. physical and chemical impacts on wetlands;
   d. additional impacts on living communities or human uses;
   e. socioeconomic effects;
   f. indirect effects;
   g. mitigation.
4. Statement of Finding: The Corps's Decision
   Permit is granted or denied.

(*Sources:* Braddock; EPA; Gaddie and Regens)

Box 5.2
Types of Permits

General:
General permits are granted for activities likely to have minimally destructive effects on wetlands. There are national, regional, and statewide general permits available for particular activities. Permit applications are waived for activities covered under general permits.

National:
Nationwide permits (NWPs) may apply to specific geographic areas, types of activities, and categories of waters. Among the activities covered under NWPs are: aids to navigation; utility line backfill and bedding; bank stabilization; and (NWP #34) Cranberry Production Activities.

Individual:
Individual permits are granted for activities likely to have significantly adverse effects on wetlands.

(*Sources:* Braddock; EPA)

Introduction to the Study of the Politics of Wetlands

Regulations to protect wetlands reflect a contemporary desire for government protection of common natural resources; they also challenge some of our most tradition political values and our most enduring political myths. All policy debates force us to decide on which side of an issue we are most comfortable erring and remind us to expect imperfections in the execution of regulations. For instance, in considering wetlands regulations, how do we balance the rights of individual citizens against **public goods**? At what point do we relax wetlands protection in deference to our strong traditions of **individualism** and the sanctity of private property? In seeking answers to these questions, the study of the politics of wetlands protection addresses issues of **public policy**, political philosophy, and constitutional law.

Public Policy Analysis

Environmental policy is not created in a vacuum: economic, historical, environmental, social, and political influences all factor into our tolerance for regulation, our respect for property, our suspicion of bureaucrats and government, our reliance on courts, and our concern for common resources. Making environmental policy is a frustrating and long-term process because it concerns our health, property, sensibilities, and future and requires thoughtful analysis of complex and unclear data. Because the stakes are high, the process attracts numerous authorized players and interested parties. As Dan Fiorino (1995) argues, environmental policymaking is messy:

> The price of democracy . . . is that people and institutions often work at cross-purposes, often toward conflicting ends, with not always satisfying results. Why else have a system in which Congress passes a law directing an agency to set strict regulatory standards, only to allow an agent of the White House to delay their issuance? Why have agencies spent years making a case for banning a chemical, only to have a judge remand for better justification? And why has Congress divided oversight authority among more than a dozen subcommittees and a score of laws, when one law makes more sense? Institutions have varied perspectives and roles; if they appear to be working at cross-purposes, it is because they often are. (22)

In addition, lobbying by **interest groups**, suits by private citizens and corporations, controversy over the U.S. Constitution, the proliferation of separate state and local programs, electoral and partisan pressures, and scientific and political disputes over delineation of wetlands all contribute to the complexity of the political ecology of wetlands policy.

Political science and the study of politics include the evaluation of public policy from many perspectives. Systems analysis is one approach to the study of public policy; it encourages the integration of analysis of groups, elites, institutions, and processes in studying public policy (Fiorino 1995). According to systems analysis, policy is the result of environmental influ-

---

Box 5.3:
Wetlands Permits Statistics 2001

Wetlands acreage impacted by Corps permits:	68,000
Number of wetlands acres filled:	25,000
Number of acres required to be restored, created, or enhanced:	43,000
Number of projects coved by General Permits:	75,847
Number of applications for individual permits received:	11, 783
Number of projects–individual permit decisions: (includes permits and denials)	7, 396
Number of permits denied: (most other permits are modified or conditioned before issuance)	171
Number of applications withdrawn:	3,791
Number of violations of permits reported: (less than 1% results in litigation)	5,170

(*Source:* U.S. Army Corps of Engineers:
http://www.usace.army.mil/inet/functions/cw/cecwo/reg/execsem01.pdf)

---

ences (such as public opinion and political philosophy) on and inputs (such as demand and support for particular policy goals) into the political system (the institutions of government). The system produces an output (a public policy) that then reenters the system, through demands and supports generated by that policy, as an input. Systems analysis concentrates on the system's response to stress and its effort to maintain stability (much like homeostasis in ecosystems). This approach to the study of policy provides a forum in which to ask, "How do characteristics of the political system affect the content of public policy? How do environmental inputs affect the content of public policy? [And how] does public policy affect . . . the environment and the character of the political system?" (Dye 1998, 36). Systems analysis offers improved understanding of the processes and interdependence of the actors, shaping environmental public policy, such as wetlands regulations. However, systems analysis is much less helpful as a tool for political evaluation of the policy in normative terms. While systems analysis might identify outputs (new requirements for wetlands mitigation) and interdependencies and influences in the political system (campaign contributions from building contractors associations or editorials on environmental groups' web pages), it does not indicate whether those outputs or influences are constitutional or just or environmentally sound. Therefore, systems analysis is criticized for its focus on the identification of outputs without evaluation of those outputs.

The following discussion of the politics of wetlands incorporates systems analysis in an overview of the history and current status of wetlands regulation. To reiterate, the *inputs* are the demands and supports for wetlands regulation. *Demands* include requests for increased wetlands protection, financial compensation for wetlands conservation, and adherence to particular political principles, such as the protection of individual rights. *Supports* uphold the selection and legitimacy of various demands. They "may be based on cultural, ideological or national loyalties. They may also rest on coercion, fear, and lack of available alternatives. Whatever their source, supports provide effective criteria—one might speak of them as natural gatekeepers—for the selection process between demands" (Susser 1992, 184). In regard to wetlands regulation, supports include understandings of property rights and environmental activism. The *political system* that selects inputs and converts them into the *outputs* of public policies consists of the government institutions and processes designed to produce laws, rules, and regulations. In this case, the executive, legislative, and judicial branches of the state and national governments constitute the political system. The *environment* in which the political system operates and in which demands are generated completes the framework of systems analysis. As discussed in this chapter, the political environment in which wetlands regulations have been created is defined by fundamental concepts and goals of American politics.

The Political Environment

The demands for wetlands regulation arise from the concept of public goods, a characteristic of the political environment. Most environmental policies are based on the notion that natural resources should be protected as public goods for the society as a whole. The politics of wetlands regulation, however, reflects intense disagreement over whether wetlands protection is in fact a public good. In considering who bears the costs and who benefits from wetlands protection, proponents of environmental regulations are trying to justify often direct and conspicuous private costs (the possible loss of economic value of land) for public goods (conservation of wetlands). The specificity of those "paying for the policy" through compliance costs, fees, and other burdens and the general dispersion of the beneficiaries of wetlands protection mean that those "paying" have a greater incentive to resist, complain, and extricate themselves from the regulatory actions. Beneficiaries of public goods, on the other hand, have fewer incentives to organize to secure their benefits (Berry 1997). So, while regulations to protect clean air and water are less controversial as public goods, regulations for wetlands protection are not accepted by their opponents as public goods because the costs are viewed as too high and the benefits as insignificant and narrow. Disputes over wetlands protection involve two views of wetlands regulations: either as policies to secure the public good of habitat protection and flood control or as obstacles to individual liberty and the protection of private property.

Perhaps no principle is as essential to American government as that of the freedom of individuals from government interference. **Liberalism** is the belief that governments ought to favor individuals and commerce by freeing both from undue regulation. Americans have inherited and embraced liberalism, although some environmental scholars have noted that liberalism is not always amenable to environmental regulation, which requires greater emphasis on public goods and public resources and therefore increased government interference in individual lives (Cahn and O'Brien 1996). As Deborah Stone (2002) writes, "Freedom is no less ambiguous and complex than other goals and values that motivate politics. . . . The dilemma of liberty surfaces in public policy around the question of when government can legitimately interfere with the choices and activities of citizens" (108–9). For many political philosophers and American citizens, individual freedom is predicated on the existence of private property, protected from arbitrary and capricious government interference.

The founders of American government derived much of their understanding of liberalism from the British political philosopher John Locke, who, in his *Second Treatise on Government*, explained the origins of private property and the relation between property and individual freedom within society. For Locke, property derives from an individual's effort and labor, whether it be in gathering fruit or cultivating fields, and the protection of property is a fundamental purpose of government. Some students of Locke argue that he was more willing to recognize regulatory limits on private property than most private property rights advocates admit. As law professor Mary Ann Glendon (1991) writes,

> From the very beginning, the absoluteness of American property rhetoric promoted illusions and impeded clear thinking about property rights and rights in general. The framers' efforts to directly and indirectly protect the interests of property owners were never meant to preclude considerable public regulation of property. In the case of property, it was . . . the Lockean paradigm, cut loose from its context, that became part of our property story as well as of our rights discourse. (25, 43)

Thus, the influence and interpretation of Locke are crucial elements of the environment surrounding the political system in which wetlands policy is made.

To review, one component of understanding how the political system responds to certain demands is that of identifying the political environment in which the system operates (Dye 1998). In the case of wetlands, we need to consider that while those pressuring the political system on behalf of private property rights argue that they are simply trying to reassert fundamental American beliefs, those advocating stricter environmental regulation argue that the demand for total autonomy over private property is a corruption of the founding principles and ignores our history of sacrificing private interests to achieve communal goals.

Several other characteristics of the political environment illustrate how historical and political attributes of the United States influence wetlands regulations. For example, environmental studies professor H. M. Jacobs (1999) contends that Americans are deeply conflicted about land and property as a result of our various interpretations of American history, the meaning of land, and the philosophy of land; he also notes that these conflicts cannot be resolved through our representative democracy. Another interpretation that contributes to understanding the political environment views the current emphasis on the individual and private property rights in the dispute over wetlands regulations as a manifestation of a weakened participatory democracy (Shutkin 2000). William Shutkin identifies several civic causes of the shortcomings of environmental regulation, including the diminution of social capital (a sense of community and participation in community activities) and the trend toward **privatization** and away from **public investment**. The theories of Jacobs and Shutkin describe a political environment less likely to favor environmental protection as a legitimate public good. Judicial opinions and legislative proposals, however, offer additional evidence that our political system has responded by establishing an uneasy equilibrium between private property protection and environmental public goods.

Two additional characteristics of the American political environment that are essential to understanding wetlands policy are **federalism** and the role of interest groups. Federalism allows the distribution of political power among the national, state, and local governments and contributes to both the political environment and the political system (in separate executive, legislative, and judicial institutions). Federalism also allows for state policies not only to differ from one another but also to vary from federal policies, within some limits. It is important to note that in most areas of environmental policy, the federal government sets standards regarding the protection of natural resources. States can exceed federal standards but cannot use less stringent state standards. The presence of interest groups in the political system can be viewed as a characteristic of the political environment, as an element of the political system, and as a source of demands on the system. Interest groups are organizations of like-minded people pursuing political and policy goals through lobbying and efforts to educate the public. One common interpretation of public policy views policy as the result of **pluralism**, that is, the result of the compromises and negotiation between interest groups. After a brief history of wetlands policy, these two influences on takings legislation are discussed next, followed by further analysis of the specific types of takings legislation.

Wetlands Regulations and the Political System

In systems analysis, policy itself is viewed as the outcome of the political system as that system responds to various pressures and attributes of the political environment. During the past 25 years, wetlands regulatory policy has frequently been described as murky, muddy, and swampy—even as a quagmire. The turbidity of wetlands regulatory policy was the focus of much political attention in the late 1980s and throughout the 1990s. Policymakers, bureaucrats charged with policy implementation, and **private property rights groups** still participate in the ongoing mutations of wetlands regulations despite the ebbing of the general public's concern with wetlands controversies.

There are a number of **federal agencies** involved with the regulation of wetlands. They include most significantly the EPA's Wetlands Division of the Office of Water and the Corps. In order to understand how wetlands regulations are developed and implemented by the EPA and the Corps, the nature of bureaucracies and the specific history and cultures of these two agencies must be considered. While bureaucracies were once idealized as apolitical, they are in fact battlefields for conflicting political and partisan interests. And all bureaucracies face constraints on resources, conflicts between various goals, and concerns about discretion and autonomy (Wilson

1989). The EPA and the Corps have different missions, different jurisdictions (or "turfs"), and different organizations and personnel; these differences shape the two agencies' approaches to wetlands regulation (Wilson 1989). The EPA consists of personnel with specialties in law, economics, and natural sciences; its mission has been interpreted variously by its chief administrators as the protection of natural resources, the protection of human health, or, occasionally, the promotion of the interests of American industries (Landy et al. 1990). The attention and resources devoted to wetlands and the ways in which the agency attempts to fulfill its mandate regarding wetlands protection will reflect the organizational characteristics of the EPA as well as the priority within the agency of national resources versus human health issues. On the other hand, the Corps has a history of destroying wetlands in the construction of **public works projects** and being oriented toward the fulfillment of engineering plans. The Corps also has a strong ethos of professionalism based on its elite military membership and its commitment to the principles of engineering. This ethos is manifest in the discretion allowed field supervisors and the Corps's clear sense of and dedication to its mission, which is the construction of public works (Wilson 1989). The **regulatory rule making** and implementation responsibilities of the Corps and the EPA in wetlands protection qualify these bureaucracies as appropriate subjects for students of politics examining wetlands policy and politics.

Wetlands policy also reflects presidential administrations and their influence on federal bureaucracies, such as the EPA and the Corps. Many aspects of a presidency, ranging from campaign strategy and management style to the exercise of specific presidential power, can illustrate the role of presidents in shaping environmental policy. Two useful examples are the presidencies of Ronald Reagan (1981–1988) and George H. W. Bush (1989–1992). Reagan favored less intrusive government and opposed environmental regulation as a hindrance to the economic competitiveness of American industries. Reagan used his presidential appointment power to place antienvironmental, antiregulatory personnel at the heads of the agencies with environmental jurisdiction, including the EPA. As expected, these appointments resulted in significant changes in agency budgets and priorities. At the EPA, strict enforcement of environmental regulations was subordinated to consideration of economic values. Reagan also influenced wetlands regulation when he issued Executive Order 12360, which instructed all federal agencies to review their rules and regulations and to evaluate whether their implementation would result in a taking of private property (Switzer 1994; Valente and Valente 1995; Vig 1994). This order had the potential to obstruct the implementation of environmental regulations that reduced property values. While this executive order was never fully implemented, it indicates the power of the president to hamper environmental regulations. In general, the business community applauded Reagan's decisions, while the environmental community and Congress were dissatisfied, fearing an evisceration of environmental regulations.

When Reagan's vice president, George H. W. Bush, sought the presidency in 1988, he campaigned as "the environmental president," clearly trying to distance himself from Reagan's record on environmental issues. In challenging his Democratic opponent, Massachusetts governor Michael Dukakis, candidate Bush promised that, if elected, he would ensure "no net loss of wetlands." Once elected, the Bush administration and Bush's supporters were greatly conflicted over how to achieve "no net loss of wetlands," how wetlands would be defined, and whether the consequences of wetlands regulations for resource extraction industries were acceptable. At the same time, the several federal agencies with jurisdiction over wetlands were collaborating on a new wetlands delineation manual designed to reflect Bush's stated wetlands protection goals. The 1989 manual proposed a broad (overly broad, in the eyes of industry) definition of wetlands. Criticism from property owners and developers immediately fell on the Bush administration, and opponents of wetlands regulations mobilized to thwart the adoption of the new manual. One response to the critics was the creation of Vice President Dan Quayle's **Council on Competitiveness.** Designed to review proposed regulations for their impact on economic competitiveness, the council represented the Bush administration's effort to now narrow the definition of wetlands (Dolan 1993). Meanwhile, President Bush, facing reelection concerns, was caught between the efforts of the federal bureaucracy to define wetlands in a way that would create more, not fewer, wetlands under regulatory protection and the efforts of his vice president and his longtime supporters in favor of less regulatory protection through a narrower definition of wetlands (Switzer 1994; Valente and Valente 1995; Vig 1994).

The presidencies of Bill Clinton (1993–2000) and George W. Bush also shaped and continue to direct federal bureaucracies and wetlands policies. Following President George H. W. Bush, President Bill Clinton echoed the former president's pledge to ensure "no net loss of wetlands." The Clinton administration, however, faced the remaining challenge of resolving the question of how wetlands were to be defined for regulatory purposes. Although President Clinton made an effort to reach a consensus about wetland delineation and protection and to initiate reforms, such as greater regulatory control over wetlands by the Department of Agriculture (Barr 1993), and despite reinstating the protection of wetlands in Alaska, he left office without any significant clarification or impact on wetlands policy.

Box 5.4
Wetlands Acreage Estimated Impacts

FY	Permitted	Mitigated
1995	26,000	46,000
1996	25,000	49,000
1997	37,000	52,000
1998	30,500	48,000
1999	21,000	46,000
2000	19,000	43,000
2001	25,000	43,000

(*Source:* U.S. Army Corps of Engineers)

The early months of George W. Bush's presidency raised concerns among many environmentalists as to the president's commitment to environmental protection, but by Earth Day in April 2001, President Bush was eager to demonstrate his environmental bona fides. One measure he took was to applaud the EPA's endorsement of the Clinton administration's last-minute rule that clarified the "discharge of dredged material" provision of Section 404 (Jehl 2001) and offered somewhat more protection to wetlands. However, some environmentalists remained skeptical of Bush's commitment to wetlands protection, given his close connections with the natural resource extraction industries (National Wildlife Federation 2001). This skepticism increased when, in 2002, the Corps, with support from the Bush White House (but facing criticism from other federal bureaucracies such as the EPA), revised its national permits to weaken wetlands protection by, among other provisions, increasing limits on the amount of seasonal streambed that can be filled under such a permit and increasing the minimal wetlands acreage that can be adversely affected under such permits (Grunwald 2001). Presidential influence on wetlands is clearly variable and a reflection of political philosophy as well as partisan and electoral concerns. The presidential positions and actions reviewed here were made in the context of a political ecology or system responding to intense pressures regarding such fundamental political values as the sanctity of private property.

Private Property Rights Demands and Judicial and Legislative Outputs

Several demands, ranging from improved implementation of wetlands protections at the state and regional levels to clearer delineation of wetlands for regulatory purposes, frame wetlands policy. One significant demand is for restitution to property owners for restrictions on land use resulting from wetlands regulations. The **Fifth Amendment** of the U.S. Constitution offers a possible balance between regulatory protection of wetlands and private property rights. In addition to protecting citizens from self-incrimination, the Fifth Amendment prohibits the government from taking private property for public use without just compensation. But what constitutes the taking of property? The complete transfer of physical property to the possession of the government is clearly a "taking," but is a partial restriction of property rights due to regulatory prohibitions? And what is public use? Despite the Fifth Amendment, government decisions may impact private property without constituting takings; however, as Supreme Court Justice Holmes said in a 1922 case, if a regulation goes "too far," it will be considered a taking (*Pennsylvania Coal Co. v. Mahon*). This extraordinary statement by Holmes marked a change in the Supreme Court's definition of takings and would became the basis of future judicial consideration of the effects of regulations on private property. Efforts to assert private property rights and to claim takings (that some or all of a piece of property has been taken from its owners as a result of compliance with wetlands regulations) have taken two main forms: judicial rulings to compensate after the fact of a regulatory taking and legislation to guarantee compensation in anticipation of regulatory takings.

Regulatory takings as opposed to **physical takings** cases are based on the Supreme Court's decision in *First English Evangelical Lutheran Church v. County of Los Angeles* (1990), which acknowledged temporary takings. Plaintiffs in regulatory takings cases argue that if temporary takings (property is affected when a regulation is in effect, then that regulation is repealed) can exist, so can takings affecting part of a property (2 acres of a 10-acre parcel), as opposed to the parcel as a whole (all 10 acres) (Emerson and Wise 1997). The Supreme Court usually decides regulatory takings cases on an ad hoc basis by examining three characteristics in order to determine if a regulation has gone "too far." The Court examines 1) the character of the governmental action, 2) the economic impact of the regulation, and 3) the interference of the regulation with investment-backed expectations (*Penn Central Transportation Co. v. New York City* 1978). The courts were besieged with takings cases regarding Section 104 permits during the mid-1990s, especially following the case of *Lucas v. South*

Carolina Coastal Council (1992). In *Lucas*, the Supreme Court determined that government regulations prohibiting all economic use of property constitute a taking and that Lucas was qualified for compensation from the government without consideration of the three criteria of ad hoc review. The *Lucas* case, like so many other aspects of wetlands politics, is open to interpretation, but most legal and environmental analysts view *Lucas* as a move toward greater protection of private property rights (Meltz 2000; Vickroy and Diskin 1991).

Despite the tradition of **stare decisis**, the courts' interpretations of takings, including those of the **U.S. Court of Federal Appeals**, have altered over the past century in response to changes in jurisprudence, the makeup of the courts, and the increased number of environmental regulations. For the most part, the state courts have been less likely to decide in favor of a taking and compensation for the property owner and more likely to decide that wetlands protections regulations, even if decreasing property values or preventing development, are necessary and legitimate (Meltz 2000). While previously the federal courts seemed to say that all uses of the parcel as a whole must be denied for a taking to exist and that public goods, such as wetlands protection, outweighed private freedoms and gains, they now seem to be more supportive of defending the individual from regulations that go "too far." The most recent Supreme Court decisions, however, are no closer to clearly defining regulatory takings than they were in the mid-1990s. Consequently, private property rights proponents, while not quite abandoning the judicial forum, turned to legislatures to pursue their goals.

During the 1990s, all 50 states saw the introduction of private property rights or "takings legislation"—legislation that would codify support for private property rights, indicate how citizens would be compensated, identify the source of funding for these compensations, and define what percentage diminution of a property's value or possible profits would qualify for compensation. While at least 26 states passed legislation, some states saw the legislation defeated outright or repealed or watered down shortly after passage (Emerson and Wise 1997; Pierce 1996; Vickroy and Diskin 1991).

Most participants in crafting takings legislation viewed regulatory takings as the result of environmental policies, such as the CWA and the **Endangered Species Act**. Some people, however, currently extend the debate over takings legislation to include its possible application to zoning regulations, provisions of the Americans with Disabilities Act, education policies, and, more recently, regulation of high-tech industries (DeLong 1999). Supporters of takings legislation are usually conservatives who champion the Fifth Amendment and private property rights while decrying the tangle of government regulations they consider inappropriately imposed on individuals and businesses. Opponents of takings legislation are usually fiscal conservatives (those concerned with government spending and balanced budgets) who view the bills as likely to desiccate state budgets and traditional liberals who fear the bills will weaken political support for needed regulations (Emerson and Wise 1997; Vickroy and Diskin 1991).

Political and legal scholars have identified a number of attributes of the political environment that fostered the introduction of takings legislation in the mid-1990s by state governments. The flood of takings legislation at the state level correlates to "the rise of . . . conflicts over land use and environmental regulation, political organizing by private property rights groups, and federal judicial pronouncements" (Emerson and Wise 1997, 411). The *Lucas* case of 1992 in particular is cited as encouraging states to satisfy the demand for greater private property rights protection in a judicial climate likely to endorse such protections. The primary objectives of takings legislation are the prevention of costly litigation against the state and the protection of private property rights for state residents without resort to legal battles (Vickroy and Diskin 1991).

State governments, encouraged by the political environment of federalism, experimented with various mechanisms to protect private property and produced a variety of takings legislation. One often cited attribute of federalism is that it allows for the incorporation of individual characteristics of local and state political culture into policy. States determine whether to consider takings legislation and what the provisions of their takings legislation will be on the basis of the previously mentioned goals above as well as on characteristics such as their political culture, natural resource wealth, major industries, and degree of statewide planning (Emerson and Wise 1997). "For example, resource-rich states (e.g., the Northwest and the Farm Belt) tend to have politically active resource extraction and agricultural industries that favor such legislation" (Emerson and Wise 1997, 413).

State legislatures were also responding to their constituents' demands for protection against federal policies, including the CWA's wetlands provisions. The federal and state governments have varying reputations in regard to environmental protection and regulatory relief, complicating the evaluation of federalism in environmental policy. The state governments took the lead in wetlands protection, passing wetlands protection policies prior to 1972, while federal policies have long contributed to the destruction of wetlands. Many thousands of the wetlands acres lost over the past decades have been destroyed as a result of federal decisions on public land use, mining, timbering, agricultural subsidies, housing, lending, and transportation. Once the federal government made the protection of wetlands a priority, it seemed to focus mostly on the actions of private property owners, as opposed to the actions of government agencies, in trying to prevent "no net loss" of wetlands. Therefore, many critics of wetlands regulations see these regulations not only as unfair to individual citizens but also as hypocritical (Miniter 1991). In addition, property owners view the layers of federal, state, and local regulations as contributors to the thicket of red tape and to the confusion over which regulations apply and which level of government ultimately controls land use decisions.

Although some examples of takings legislation indicate a desire of state legislatures to offer symbolic support for property rights and make few changes to existing policies, most state takings legislation falls into two categories: assessment and compensation legislation (Emerson and Wise 1997; Oswald 2000; Vickroy and Diskin 1991). Assessment legislation requires state and sometimes local governments to prepare a takings impact statement for regulations to determine whether the consequences of the regulations would be regarded as regulatory takings. These "look before you leap" bills were designed to allow states to identify likely costs of compensation for regulatory takings before implementation and to save property owners the trouble of establishing that compliance with a particular regulation constitutes a taking should they qualify for compensation (Emerson and Wise 1997). These bills were modeled on President Reagan's 1988 Executive Order 12630, which was never fully operational within the federal government; analysts suspect that these efforts are likely to face similar implementation problems at the state level because they require definitions of regulatory takings, which thus far neither the federal government nor the Supreme Court has been able to articulate (Vickroy and Diskin 1991). Compensation provisions are designed to compensate property owners for a certain percentage decrease in their property's value as a result of state-level regulation. The percentage at which property owners will be compensated for devaluation of their property varies from state to state. Several states define regulatory taking as a 50% reduction in property value, while others define takings at reductions of 40% or 25%. Some state proposals considered takings as occurring at as low as a 10% devaluation. Many compensation bills also identify certain government actions that are noncompensable. For example, Texas excludes groundwater management and floodplains regulations from compensation in its takings bill (Vickroy and Diskin 1991).

Battles in some states over the passage of takings legislation have been fierce and have attracted national participants, reflecting the intensity of the battle over interpretation of the Fifth Amendment in the protection of the environment. In Washington and Arizona, national interest groups, such as the Sierra Club, joined the effort to determine the outcome and provisions of takings legislation. Both states also saw the defeat or repeal of stringent takings bills by a margin of 60% to 40% of the vote. Currently, the boldest takings legislation is that of Florida, which passed its Private Property Rights Protection Act in 1995, extending compensation to case of regulations that impose an "inordinate burden" on property owners (Carreja 1996; Emerson and Wise 1997; Vickroy and Diskin 1991). A property owner carrying an unfair share of the costs of providing a public good is considered to bear an "inordinate burden." Legal scholars argue that Florida is not simply indicating its support for the Fifth Amendment but is seeking to broaden the protection of property rights (Carreja 1996; Emerson and Wise 1997; Vickroy and Diskin 1991). Opponents of Florida's takings legislation fear the bill will bankrupt the state and prevent the passage and implementation of needed regulations to protect public welfare. Further variation of state takings legislation exists as some states' versions apply to particular types of property, as in Louisiana and Mississippi, where only forests and agricultural lands are covered (Vickroy and Diskin 1991). Therefore, despite national participants and the boilerplate versions of takings legislation available from organizations like the American Legislative Exchange Council and Defenders of Property Rights (Emerson and Wise 1997), state takings legislation is not uniform, and neither are these legislative efforts uniformly successful in defining or compensating takings.

A number of states have witnessed the defeat of takings legislation, indicating that the public is wary of codified means of compensation as opposed to reliance on the courts to resolve questions regarding takings. It is curious that takings legislation has not been more successful given the assumed American devotion to individualism and private property rights. Neal Pierce (1996) argues that the costs and bureaucracy involved in compensation are off-putting to Americans concerned about budget solvency and desiring smaller government. It may also be that unless they are directly involved in a case of regulatory taking, most Americans still trust the legitimacy of regulatory policy and believe in the common goals and public goods such policy is supposed to secure.

Takings legislation manifests the efforts of groups representing both environmental and private property rights interests to shape wetlands policies. These interest groups influenced takings legislation through testimony before legislative committees, through endeavors to define policy by participating in court cases, and through efforts to shape public opinion regarding wetlands and private property rights. Environmental interest groups, such as the National Wildlife Federation and the Sierra Club, lobbied members of Congress and state legislatures to ensure continued wetlands protection and to oppose takings legislation. Some environmental interest groups, such as the Natural Resources Defense Council, were crucial to the expansion of wetlands protection by filing suits to push the courts to determine the authority of the EPA and the Corps in regard to wetlands. However, environmental groups have not been so successful in countering the private property rights movement or in explaining and educating the public about the tensions between environmental goals and personal property rights (Dowie 1996).

Equally interesting for the political analysis of wetlands is the role of the private property rights movement, often identified as part of the **Wise Use movement**. One measure of the success of the private property rights interest groups, including such groups as the Alliance for America and the League for Private Property Rights, is that they have "redirected public discourse about land and environmental policy; more and more legislators, members of the media, and members of the American public now view the issue of land use and environmental policy from the perspective of how it impinges on private property, rather than how it furthers public goals" (Jacobs 1999, 22). Wise Use groups seek to open public lands to increased recreational and economic uses, to reduce the burdens of environmental regulations, and to return to the principles of the early American republic as they see them. Jon Roush (1995) reminds us that "the two motives of the Wise Use movement—defending private property and the private use of public land—rest on venerable American values and institutions" (5). While some commentators argue that the Wise Use movement is simply a front for industry, especially natural resource industries, many members of the movement argue that the federal government violates the Constitution through environmental regulations, abusing the rights of individual citizens. Because of its controversial nature, some private property rights groups resist being identified with the Wise Use movement.

Supporters of weaker wetlands regulations and stronger takings legislation also include representatives of developers and contractors, such as the National Association of Home Builders; representatives of agricultural businesses, such as the Farm Bureau; representatives of natural resource extraction industries, such as the National Coal Association; and coalitions of contractors, developers, and industries, such as the National Wetlands Coalition. This last group, despite its name, is interested not in wetlands protection but rather in maintaining its members' ability to develop wetlands with the fewest regulatory inconveniences and obstacles. These groups, like the environmental groups, lobby legislators and file lawsuits, in this case to establish takings and restrict the jurisdiction of the EPA and the Corps.

Takings legislation at the federal level has faced similar obstacles created by interest group pressures and concerns about budgetary impacts, but there have been some successes since the Republican Party resumed control of Congress in 1995. One element of the 1994 Republican congressional platform, Contract with America, was a takings bill that would compensate owners for loss of 10% or more of property value due to laws protecting wetlands and endangered species; it passed the House of Representatives easily with the devaluation limit increased to 20%. However, by the spring of 1995, when similar bills were considered in the Senate, editorial pages across the country were urging caution and opposition. In 1998, federal takings legislation was again the focus of debate, although with somewhat less salient controversy; this bill, supported strongly by the National Association of Home Builders, created what proponents called a more streamlined process to determine regulatory takings and secure compensation by allowing challenges to permit decisions earlier, something opponents saw as an effort to bypass regulatory authorities and the appropriate judicial process. Federal efforts to pass takings legislation seem to have the most success in the House; this may be attributed to one interpretation of the nature of the House as a more immediate conduit of public desires and whims and less likely to emphasize the fiscal and political consequences of policies.

The Future of Wetlands Protection, Regulations, and Takings Legislation

As an outcome of the political system, takings legislation offers a mixed bag of results: While no major federal takings legislation has become law, several states have passed legislation that has been in effect only for a few years, making evaluation of the legislation difficult. In addition, most takings legislation contains provisions that create more discretion for state and local bureaucracies and courts, an ironic outcome of legislation promoted by opponents of bureaucracy in general and those dissatisfied with the judicial protection of private property rights offered through the Fifth Amendment.

Federalism, bureaucratic complexities, judicial decisions, and interest groups all contribute to the mutability of wetlands regulations and the difficulty in fashioning a necessary, feasible complement to the Fifth Amendment. As with so many areas of public policy, when it comes to wetlands protection and private property rights, Americans want to have their cake and eat it, too—to have their own property rights protected while those of others are limited to ensure some public good. Jacobs (1999) identifies three goals of Americans regarding private property land use policies. He notes that we want security for our financial investments in real estate, we want clarity in the rules and regulations regarding land use, and we also want certainty regarding the effect of any changes to regulations on our security. The controversies over regulatory takings and the general displeasure with Section 404, despite the efforts of the courts and states legislatures, indicate how far we are from securing these goals.

In conclusion, there are a number of factors likely to impact the future wetlands regulations and takings issues and the protection of wetlands in general. The decisions of the state and federal courts regarding wetlands regulations and takings reflect not only jurisprudence but also electoral outcomes. Congressional and presidential elections may have an indirect impact on wetlands policy through appointments to the Supreme Court and the other federal courts. Still open to evaluation is the fact that the filing of takings cases regarding wetlands regulations has subsided. Regardless of the trend of fewer takings cases, the Supreme Court may be called on in the near future to resolve questions left unanswered by their previous decisions.

On the legislative front, reforms to the CWA have been introduced to weaken the jurisdiction of EPA and the Corps over wetlands. Future consideration of the reauthorization of the CWA may also include specific provisions dealing with and defining wetlands protection. Changes to federal agricultural policies will affect wetlands regulation; most significant is the 2002 reauthorization of the Farm Bill, which garnered the attention of both sides of the wetlands controversy. The 2002 Farm Bill's Section 2201 restored the Wetlands Reserve Program and increased the wetlands acreage eligible from protection under this program. This increased protection of wetlands may be offset by changes to energy and resources extraction policies. Finally, as the experiments in other areas of environmental protection continue, the spectrum of nonregulatory measures to secure wetlands protection may include privatization and market-based incentives.

The Bush administration supports increased resource extraction on public lands, including the Arctic National Wildlife Refuge; this position will negatively affect wetlands on public lands. In addition, President Bush is likely to appoint (and a closely split Senate is likely to confirm) federal judges and Supreme Court justices who share his perspective on environmental issues, including land use and property rights. With heightened threats to the environment, it is probable that the Bush administration, as did the Reagan administration, will jump-start the flagging environmental movement. The rejuvenation of the environmental movement will succeed only if its leaders take the private property rights activists seriously, are sensitive to manifest hostility to environmental regulations, and respond to public skepticism about the effectiveness of rigorous restrictions on private activities (Dowie 1996).

However, the future of wetlands regulations seems most likely to be one of continued debate over and mutation of wetlands delineation contributing to bureaucratic inconsistency and citizen frustration—a future in which decisions in regulatory takings cases are ad hoc and not very predictable and in which takings legislation is part of the agenda of "antigovernment" candidates, elected officials, and interest groups. While takings legislation continues to be introduced in both federal and state legislative bodies, the consideration of such legislation seems to be mostly symbolic. It is unlikely for both procedural and fiscal reasons that any such legislation will drastically reform the regulatory nature of environmental policies. Public opinion does not seem likely to swing significantly on wetlands or private property issues, although votes repealing or defeating takings legislation have resulted in margins narrow enough that future considerations might be affected by motivated turnout and a particular crisis or issue.

Garrett Hardin (2002), in "The Tragedy of the Commons," argues that environmental resources such as wetlands require protection by "mutual coercion mutually agreed on by a majority of the people affected" (26). Such protection assumes regulatory policies that are viewed as legitimate (framed by representatives of the people, shaped by consensus, and affirmed as constitutional by the courts). The more consistent and knowable such policies are to both private citizens and civil servants without depending on increasingly rigid, complex, and top-down regulations, the better off wetlands and public attitudes toward government will be. Systems analysis of wetlands policy demonstrates why this remains the political challenge for wetlands protection.

REFERENCES

Barr, Stephen. 1993. Clinton to revise wetlands policy; plan tries to satisfy conflicting groups. *Washington Post*, August 25, A1.

Berry, Jeffrey M. 1997. *The interest group society*. 3rd ed. New York: Longman.

Braddrock, Theda. 1995. *Wetlands: An introduction to ecology, the law, and permitting*. Rockville, Md.: Government Institutes, Inc.

Cahn, Matthew Alan, and Rory O'Brien. 1996. *Thinking about the environment: Readings on politics, property and the real world*. Armonk, N.Y.: M. E. Sharpe.

Carreja, Tirso M., Jr. 1996. Adding a statutory stick to the bundle of rights: Florida's ability to regulate wetlands under current takings jurisprudence and under the Private Property Rights Act of 1995. *Journal of Land Use and Environmental Law* 11, no. 2:423–66. www.law.fsu.edu/journals/landuse/vol112/carreja.html.

DeLong, James V. 1999. Taking back the Fifth. *Reason* 31(2):32–37.

Dolan, Maura. 1993. EPA to adopt compromise on wetlands development: Environmentalists declare new regulations a victory. Quayle fails to reduce protected acreage. *Los Angeles Times*, January 14, A3.

Dowie, Mark. 1996. *Losing ground: American environmentalism at the close of the twentieth century.* Cambridge, Mass.: MIT Press.

Dye, Thomas R. 1998. *Understanding public policy.* 9th ed. Upper Saddle River, N.J.: Prentice Hall.

Emerson, Kirk, and Charles R. Wise. 1997. Statutory approaches to regulatory takings: State property rights legislation issues and implications for public administration. *Public Administration Review* 57(September/October):411–22.

Fiorino, Daniel J. 1995. *Making environmental policy.* Berkeley: University of California Press.

First English Evangelical Lutheran Church v. County of Los Angeles. 1990. 110 S.Ct. 866.

Gaddie, Ronald Keith, and James L. Regens. 2000. *Regulating wetlands protection: Environmental federalism and the states.* Albany: State University of New York Press.

Glendon, Mary Ann. 1991. *Rights talk: The impoverishment of political discourse.* New York: Free Press.

Grunwald, Michael. 2001. Army Corps seeks to relax wetlands rules. *Washington Post*, June 4, A1.

Hardin, Garrett. 2002. Freedom in a commons brings ruin to all (excerpt from "The Tragedy of the Commons"]. In *The state and nature: Voices heard, voices unheard in America's environmental dialogue.* Edited by Jeanne Nienaber Clarke and Hanna J. Cortner. Upper Saddle River, N.J.: Prentice Hall, 242–48.

Jacobs, Harvey M. 1999. Fighting over land: America's legacy . . . America's future? *Journal of the American Planning Association* 65(2):141–49.

Jehl, Douglas. 2001. E.P.A. supports protections Clinton issued for wetlands. *New York Times*, April 17, A1.

Landy, Marc K., Marc J. Roberts, and Stephen R. Thomas. 1990. *The Environmental Protection Agency: Asking the wrong questions.* New York: Oxford University Press.

Lucas v. South Carolina Coastal Council. 1992. 112 S.Ct. 2886.

Meltz, Robert. 2000. Wetlands regulation and the law of property rights "takings." Congressional Research Service Report for Congress, RL30423. February 17. www.cnie.org/nle/wet-6.html.

Miniter, Richard. 1991. Muddy waters: The quagmire of wetlands regulation. *Policy Review* 56(spring):70–81.

National Wildlife Federation. 2001. Bush administration lets wetlands rule take effect. April 17. www.nwf.org/wetlands.bushepa.html.

Oswald, Lynda J. 2000. Property rights legislation and the police power. *American Business Law Journal* 37(spring):527–62.

Penn Central Transportation Co. v. New York City. 1978. 99 S.Ct. 226.

Pennsylvania Coal Co. v. Mahon. 1922. 43 S.Ct. 158.

Pierce, Neal R. 1996. Takings—The comings and goings. *National Journal*, January 6, 37.

Roush, Jon. 1995. Freedom and responsibility: What we can learn from the Wise Use movement. In *Let the people judge: Wise use and the private property rights movement*. Edited by John Echeverria and Raymond Booth Eby. Washington, D.C.: Island Press, 1–10.

Shutkin, William. 2000. *The land that could be: Environmentalism and democracy in the twenty-first century*. Cambridge, Mass.: MIT Press.

Stone, Deborah. 2002. *Policy paradox*. Rev. ed. New York: Norton.

Susser, Bernard. 1992. *Approaches to the study of politics*. New York: Longman.

Switzer, Jacqueline Vaughn. 1994. *Environmental politics: Domestic and global dimensions*. New York: St. Martin's.

U.S. Environmental Protection Agency. 2001. *Wetlands Definitions*. www.epa.gov/owow/wetlands/what/definitions.html.

Valente, Christina M., and William D. Valente. 1995. *Introduction to environmental law and policy*. St. Paul, Minn.: West.

Vickroy, Frank A., and Barry A. Diskin. 1991. Advances in private property protection rights: The states in the vanguard. *American Business Law Journal* 34(summer):561–605.

Vig, Norman J. 1994. Presidential leadership and the environment: From Reagan to Bush to Clinton. In *Environmental Policy in the 1990s*. 2nd ed. Edited by Michael E. Kraft and Norman J. Vig. Washington, D.C.: Congressional Quarterly Press, 95–118.

Wilson, James Q. 1989. *Bureaucracy*. New York: Basic.

S U G G E S T E D   R E A D I N G S

Emerson, Kirk, and Charles R. Wise. 1997. Statutory approaches to regulatory takings: State property rights legislation issues and implications for public administration. *Public Administration Review* 57(September/October):411–22.

Gaddie, Ronald Keith, and James L. Regens. 2000. *Regulating wetlands protection: Environmental federalism and the states.* Albany: State University of New York Press.

Miniter, Richard. 1991. Muddy waters: The quagmire of wetlands regulation. *Policy Review* 56(spring):70–81.

The Value of Resources

AN ECONOMIC
PERSPECTIVE ON
WETLANDS

Thomas Michael Power

At its most basic level, **economics** is the study of how scarce resources are managed to satisfy people's needs and desires. Economists measure the success of the economy by how productive it is in squeezing out as much human satisfaction as possible given the constraints of scarcity.

Note that this definition of economics says nothing about commercial businesses, markets, or money. Economics focuses on guiding scarce resources toward the satisfaction of human preferences regardless of whether commercial markets control those resources or are used to satisfy those preferences. Note also that this definition does not pick and choose among human needs and desires. All human desires are relevant to economics if their satisfaction requires the use of scarce resources that have valuable alternative uses.

In the past, wetlands have been looked at as "wastelands" because periodic standing water and waterlogged soils prevented their use for agriculture, forestry, or mineral production; for human habitation; or for commercial use. Before these lands could be used for one of these purposes,
they had to be drained, and drained they were: In half of the states, 50%
to 90% of wetlands have been lost; in almost all the states, a quarter or
more of wetlands have been lost (National Research Council 1992).

More recently, we have discovered that this destruction of natural wetlands
has triggered a whole range of unpleasant and unproductive consequences,
ranging from periodic flooding to deteriorating water quality to loss of
wildlife. We now recognize that wetlands are not wastelands but scarce resources that have multiple alternative uses, each of which can satisfy some
human needs and desires. The columns of figure 6.1 list, in turn, some of
these ecological functions of wetland, the ecological effects of changes in
those wetlands, and the societal values that are impacted. The management
of wetlands is, among other things, a classic economic problem: What mix of
uses of a particular wetland and protection of its natural functions is most
appropriate given people's diverse needs and desires? This chapter explores
how economists have gone about trying to answer that question.

The Role of Economics in Wetland Management

Economics seeks to provide a conceptual structure in which to evaluate
the consequences of decisions to modify or protect wetland natural systems in the pursuit of various human values. Two flows of valuable wetland services can be modified by human activity: the flow of goods and
services directly from the natural wetland systems and the flow of goods
and services from human-controlled systems that rely on those natural
wetland systems. For instance, the destruction of wetlands may reduce
the water absorption capacity to the land, accelerating water runoff and
contributing to flooding. That flooding, in turn, may destroy valuable
property and reduce economic output. One might be tempted to label the

Figure 6.1. Moving from Ecological Functions to Economic Values in Wetland Management, Selected Examples

ECOLOGICAL FUNCTION ECOSYSTEM CAPABILITIES	ECOLOGICAL EFFECT ANNUAL ECOSYSTEM GOODS & SERVICES	SOCIETAL ECONOMIC VALUES		
		INTERMEDIATE GOODS & SERVICES	FINAL GOODS & SERVICES	FUTURE GOODS & SERVICES
Hydrological Short-term surface water storage; Long-term surface water storage; Maintenance of high water table	*Hydrological* Reduced downstream flood peaks; Maintenance of base flows & seasonal flow distribution; Maintenance of hydrophytic communities	Flood Control; Water Storage; Irrigation & Sub-Irrigation water for agriculture	Flood damage security; Reduced household utility costs; Maintain sport fishing habitat in dry periods.	Unique species, landscapes & ecosystems; Bequest value; Option value; Undiscovered goods.
Biogeochemical Transformation, cycling of elements; Retention, removal of dissolved substances; accumulation of peat; Accumulaton of inorganic sediments.	*Biogeochemical* Maintenance of nutrient stocks; Reduced transport of nutrients downstream; Retention of nutrients, metals, other; Retention of sediments and some nutrients.	Assimilation of wastes; Pollution assimilation/water purification	Higher water quality as an amenity	Unique species, landscapes & ecosystems; Bequest values; Option value; Undiscovered goods.
Habitat & Food Web Maintenance of characteristic plant communities; Maintenance of characteristic energy flow	*Habitat & Food Web* Food, nesting, cover for animals; Support for populations of vertebraes	Support for commercial fisheries and recreation; provision of commercially harvested natural resources (timber, fur-bearers, etc.)	Outdoor recreation: Fishing, hunting, camping, hiking, boating, bird watching, etc.; Scenic beauty, diversified landscapes; Educational value; Existence value	Unique species, landscapes & ecosystems; Option value; Bequest value; Undiscovered goods.

first "environmental impacts" and the second "economic impacts." Such a simple division may be useful at times, but it may also obscure the fact that the natural system goods and services (such as flood control, water quality, and wildlife) contribute to human well-being in exactly the same way that the output of human activities (such as food, housing, and utilities) do. Environmental services are also "economic" in character.

In evaluating a proposed modification of how a wetland system is managed, one necessarily has to weigh the positive and negative changes that result and come to an overall conclusion about the private or social rationality of the proposal. That is, the competing and partially offsetting changes in the various goods and services of importance to human beings have to be evaluated side by side so that a rational choice about the most beneficial course of action can be made. Economics seeks to provide a conceptual structure for that comparison of the positive and negative changes, a systematic analysis of the "benefits" and "costs." This does not necessarily mean that only dollar values matter or that everything has to be expressed in dollar values. The conceptual structure provided by economics is much broader than that, focusing primarily on the analysis of trade-offs, however expressed, rather than only or primarily on dollar values and commercial outcomes. In many policy settings, however, where commercial gains or losses are at issue, it is useful, if it can be done accurately, to express environmental values in the same dollar terms. If a "common denominator" can be used without distortion or loss of information, it can facilitate decision making.

The prerequisite for economic analysis of wetland management is the tracing of environmental change to an impact that citizens can understand and/or experience. Some have labeled this the "indirect" economic value of an environmental resource since what is being valued is a particular consequence of the environmental change rather than the environmental resource itself. In order to be able to do this, economists have to work closely with environmental scientists to understand the human consequences of environmental change. A scientifically accurate description of a particular environmental change will be of little use in economic valuation if those changes are not linked to the needs and desires of humans (including the human need and desire to preserve various natural systems). If economic analysis is intended, economists should be involved at the beginning in the planning of the environmental assessment. Figure 6.1 outlines the way in which scientific descriptions of the change in the ecological function of a wetland (column 1) could affect the flow of ecological goods and services (column 2) that in turn can have an impact on a broad range of human values that can be given economic expression (columns 3–5). The scientific fact that wetlands hold precipitation and release it slowly to rivers and streams may have little economic meaning until its implications, for instance, to the threat of flooding are made clear.

Economic analysis begins with people's understanding of the way in which a particular change will influence their well-being. People cannot evaluate environmental services if they are not aware of how changes in the availability of those services will affect them. For many wetland ecosystem services, the public may simply be unaware of both the science involved and the human implications of a change. In that setting, they cannot contribute to economic valuation. That, however, does not mean that economic valuation is not possible or that the valuation of wetland management alternatives has to be left to ecologists and other scientists who are "better informed." Such experts may be better informed about the physical and biological processes involved, but there is no reason to believe that they are better informed about the relative importance of the diverse impacts on human well-being at issue.

The Range of Economic Consequences

There are a variety of economic consequences of wetland management that are important to people. These economic consequences can be divided into three categories:

1. The change in the value of the goods and services citizens receive as measured by the contribution those economic outputs make in satisfying people's needs and desires. These consequences can be labeled **economic values**.

2. The change in the level of economic activity within the local economy, especially the impact on the level of local employment and income. These consequences can be labeled local **economic impacts.**
3. The change in the distribution of costs and benefits within the economy, especially changes affecting groups of special public concern,
such as low-income households, family farm operators, Native
Americans, and future generations. These consequences can be discussed under the label **economic equity.**

Different groups are likely to emphasize different economic consequences
in evaluating wetland management. Economists and those concerned with
the efficiency of the overall economy are likely to emphasize the production of net economic value. Local economic development organizations,
chambers of commerce, and labor unions are likely to emphasize local
economic impacts. Those concerned about fairness or the disadvantaged
groups within society are likely to focus on economic equity. A complete
economic analysis of wetlands management should seek to speak to all the
economic consequences of concern to citizens.

Evaluating the Economic Consequences of Wetland Management

Figure 6.2 categorizes some of the economic consequences that might be
associated with a change in the management of a wetland to enhance its
value, such as agricultural productivity. It provides a road map to the questions that need to be explored, at least informally. Note that there are two
dimensions to this outline of the potential impacts of wetland management. The vertical dimension seeks to take account of the full range of
economic consequences outlined previously.

Some of the economic consequences of wetland management decisions
will be directly reflected in commercial markets, as different businesses
have more or less to sell. Other impacts will be noncommercial in charac-

Figure 6.2. Analysis of the Economic Consequences of Wetland Management Decisions

Accurate Depiction of the *Ecological Changes*	*Economic Consequences*	Impact on Market-Oriented Economic Activity		Impact on Nonmarket Value	
		Direct Impact on the "Intended" Industry	*External Impacts* on Other Industries Reliant on Ecosystems	*External Impacts* on *Households'* Enjoyment of Ecosystems	*Intrinsic Values* Impacts Not Associated with Use
Impact on *Ecosystem Services* of Recognized Human Concern	*National Economic Value* Perspective	Changes in Net Value of *Farm Output* (Farm output, flood damage to farm land and buildings)	Changes in the Value of *Nonfarm* Output (Flood Damage to Property and Business; Recreation; related firms; Subsidy Effects)	*Use Values* Non-commercial Recreation; Wildlife; Ecosystem Services (flood control, water quality)	*Nonuse Values* Damage to Unique Natural Systems; Endangered Species Concerns

(continued)

Figure 6.2. (continued)

Perspective	Competing Impacts within Commercial Sectors		*Amenity Effects*	Impact of *Legal Constraints*
	Gains in "Intended" Industry	Losses in "Unintended" Industries	*Amenity Effects* Quality of Life; Attractiveness of the Region to New Residents and Visitors; Local Expenditures	Impact of *Legal Constraints* Imposed to Protect National Concerns
Local *Economic Impacts* Perspective				
Equity Perspective	*Dynamic Analysis of the Local Economy:* Expected long-term Impacts including Ripple Effects and Offsets			
	The Distribution of Benefits and Costs among the Affected Parties: Farm Operators, Local Residents, Low Income, Minority Groups, Future Generations			
	Effects on the Structure of Rights and Responsibilities			

ter but may have very real impacts on individuals' well-being. This is shown in the horizontal dimension of figure 6.2.

A further distinction can be made between those impacts that take place in a market setting between willing buyers and sellers and **external impacts** (**externalities**), or those that are experienced outside of market exchanges. The additional crops that could be grown and sold if a wetland were drained constitute an example of a direct market impact. Increased flooding that impacts downstream landowners and businesses is an example of an external impact on economic values. Some of these external impacts are evaluated by markets, such as flood damage to buildings. Other external impacts affect goods and services that are not bought and sold in markets, such as water quality and wildlife habitat. Economic analysis specifically seeks to take into account both direct and external impacts as well as the market and nonmarket impacts.

Looking More Closely at Economic Values

Economic value is the capacity of goods, services, and resources to make a positive contribution to people's well-being. The conceptual measure of that capacity is the sacrifice people are willing to make of other valued things in order to gain or retain access to the goods or services at issue. This can be measured either in terms of what they are willing to give up or in terms of what they would have to be provided with before they would voluntarily relinquish their claims. In conventional economic analysis, this leads to the "willingness to pay" or "willingness to accept compensation" measures of economic value, although those concepts do not have to be expressed in money terms; anything of value, expressed in barter terms, will do. Additional farmland, reduced flood damage, improved water quality, expanded outdoor recreation opportunities, and a more varied and pleasing landscape all involve changes in this type of economic value.

There are several aspects of economic value that need to be emphasized. First, the focus is on benefits and costs of wetland management to whoever experiences them. The analysis seeks a full accounting of all benefits and costs. Local residents' values count the same as nonresidents', farmers' values count the same as urban dwellers', and so on. Second, the focus is not only on monetary benefits and costs that result from market transactions affected by wetland management decisions. If individuals' well-being is either enhanced or damaged because of a change in the way wetland resources are used, the positive changes are treated as benefits and the negative changes as costs regardless of whether those changes are expressed in money terms by the market economy. Changes in sport fishing potential are evaluated just as is flood damage or reduced crop production. Third, labor costs are treated as economic costs just as is the use of any other input to the production process. This is in contrast to local economic impact analysis that treats labor employment and payroll as two of the primary benefits of local economic activity.

This last point is not just an academic fine point promoted by economists interested in analytical purity. Economists insist on this treatment of labor costs in analyzing the production of economic value in order to keep the entire economy from being treated as a gigantic government "make-work" project with all the attendant inefficiencies and waste. Only by focusing attention on increasing the value obtained from our scarce resources can efficiency and productivity be assured. This is of considerable practical significance to all citizens and economic actors, including those concerned about environmental quality. A market economy can be trusted to use scarce resources well only if the prices that guide economic decisions accurately reflect both cost and value. If they do not, enormous waste is possible, waste that can significantly reduce everyone's productivity and prosperity below what it could otherwise have been. Such waste also, of course, can burden the environment, as more resources have to be extracted and used than otherwise would be the case if the economy were more efficient. Government-subsidized water projects have done enormous environmental damage around the world while simultaneously wasting commercially valuable resources. This doubly damages human well-being.

Economists have developed a variety of tools to measure many of these nonmarket economic values in dollar terms. Some of those tools are built around studying the actual choices people make as they pursue access to environmental goods and services. Travel costs, higher property values and cost of living, and lower wage levels, for instance, act as "entrance fees" for the enjoyment of site-specific environmental qualities. They are the sacrifices that people have to make in order to enjoy those particular sites. The study of these sacrifices can provide important quantitative evidence on people's "willingness to pay" to enjoy higher-quality living environments and, therefore, evidence of their economic value. The studies of property values and wage differentials are labeled "hedonic" analyses since they seek to relate the direct and indirect prices people pay for access to the particular attractive qualities found locally. For instance, statistical analysis of how property values vary with the level of air pollution or crime rates can indicate the economic value of clean air or lower risk of crime to people.

For many important environmental values, active use of a particular natural landscape is not necessary. For instance, people may be willing to make sacrifices to see that the grizzly bear, wolf, or lynx survive in their natural habitats, but those same people may also have no interest in sharing that habitat with these carnivores. They value the survival of these wildlife species, but this is not a use value. They may be valuing the mere continued existence of these species (existence values) or the protection of the opportunity that they might witness this wildlife in the wild in the future (option value), or they may want to make sure these members of the natural world remain available for future generations to enjoy (bequest value).

Nonuse or, as some label them, passive use values can be evaluated only through survey work that asks people about the relative importance of these environmental characteristics to them. Economists explore the trade-offs people are willing to make in the pursuit of these types of value by phrasing their survey questions in a form familiar to all of us as a result of the many economic decisions we have to make daily. For instance, if a proposed hydroelectric facility would eliminate extensive wetlands that are an important habitat for migrating waterfowl, the economist might ask people whether they would be willing to pay a somewhat higher monthly electric bill if doing so would protect this waterfowl habitat. By varying the dollar amount they question people about, information on people's willingness to pay to protect this habitat can be developed. These survey techniques that mimic common economic decisions are called **contingent valuation**. These survey results can be tested for reliability by making sure they are reproducible in many slightly different settings; by comparing the results, where possible, to market-related behavioral analysis of the sort mentioned previously; and by comparing the resulting estimated nonmarket economic values to market expenditures people make in the pursuit of commercial recreation or commercially created attractive environments.

The contingent valuation approach to estimating nonmarket economic values has been widely accepted and used by economists. The U.S. Water Resources Council has approved its use in federal benefit-cost analysis. In recent years, as contingent valuation analysis came to be used to help establish the natural resource damages that corporations have to pay for past pollution, it has come under considerable industry attack and more skeptical and critical review by some economists. In general, however, contingent valuation analysis continues to enjoy the support of both economists and environmental agencies. It is often the only tool available to monetize many environmental economic values (National Oceanic and Atmospheric Administration 1993; U.S. Water Resources Council 1983).

Local Economic Impacts

Wetland management decisions can have significant impacts on local employment, income, and business activity. Changes in wetland status can trigger changes in revenue flows through a community. Draining wetlands in the pursuit of commercial objectives may lead to increased jobs and income. Improved recreation opportunities and wildlife habitat may lead to increased fee hunting and increased recreational spending in the local area. Improvements in the aesthetics of a regions towns and cities and reduced costs of living and operating there (for instance, flood damage) may attract new population and businesses. Many, if not most, residents would see such increases in local economic activity as an improvement in the local economy, especially if that expansion is modest in scope and does not threaten traditional community values. Of course, there may also be some offsetting negative impacts associated with wetland restoration, such as loss of income associated with reduced crop production, that should also be taken into account.

In many communities, the economic dialogue about proposed changes in wetland management focuses almost exclusively on how local businesses, employment, income, and tax base will be affected by any change in wetland policy. The local economic impacts of wetland management decisions on, say, the level of employment may be only distantly related to the change in the economic value of the wetland services, such as flood control or water quality. Changes in the economic value of environmental goods and services and changes in the local level of economic activity clearly involve two quite different economic consequences of wetland management.

The phrase "local economic impacts" has come to mean an estimate of the effect on the local level of employment, income, tax revenues and public expenditures, population, and gross dollar volume of business. The estimate of changes in economic value produced may overlap local economic impacts because changes in the value of the output of local farms, for instance, are considered in both of these economic consequences. But impacts on nonlocal business activities will also be included in the economic value measure, as will changes in the noncommercial goods and services that flow from wetland systems. The economic value calculation can be expected to diverge significantly from the local economic impact measure (Courant 1994). That is the reason that the U.S. Water Resources Council distinguishes a "national economic development" accounting system (economic values) from a "regional economic development" (local economic impacts) accounting framework (U.S. Water Resources Council 1979).

There is, however, another important (but incomplete) link between local environmental quality (economic value) and local economic vitality
(economic impact) that may lead them to move somewhat together. If the
wetland values that both ecologists and economists discuss are significant
and have tangible effects on people and communities, one would expect
that changes in these would affect the attractiveness of a particular location as a place to live, work, and do business. If people care where they live
and businesses also care where people prefer to live because that determines where labor supply and markets are located, environmental quality
will affect the location of population and economic activity. People and
economic activity will be drawn toward higher-quality living environments. Note that this movement of people and businesses toward what are
perceived to be higher-quality living environments is a quite different set
of economic forces than the more popular assumption that people simply
move to where jobs are. Both are important determinants of the location
of economic activity. Neither should be ignored.

Economic Equity

Wetland management decisions that generate benefits significantly in excess of costs and that overall are judged to have very positive local economic impacts may still be criticized because the distribution of those economic consequences is judged to be inequitable. A wetland restoration
effort, might, for instance, reduce agricultural production in a rural area
while enhancing downstream property values in urban areas and providing outdoor recreation opportunities primarily for urban residents. Equity
concerns also arise when the costs associated with one group's actions are
shifted onto others, such as when the degradation of riparian wetlands
passes increased flood risks on to those downstream.

Widely shared social concerns suggest that distribution of economic
impacts among the following groups is likely to be important in wetland
management decisions:

1. Low-income and minority households: environmental justice considerations
2. Family farms
3. Existing residents as opposed to potential in-migrants
4. Future generations as opposed to current residents
5. Impacts on citizens' rights (including property rights) and responsibilities, such as the distributions of costs and benefits among cost causers and benefit recipients

Although there is no objective standard against which to unambiguously judge equity, there are widely shared social concerns that allow the economic analyst to contribute to equity discussions of the distribution of economic consequences. At the very least, a full description of the equity aspects of wetland choices ensures that any inequitable consequences were weighed in the decision-making process rather than being ignored as a result of oversight. Being aware of potential inequitable consequences early on in the planning of restoration projects also increases the likelihood that a project can be modified to reduce the potential inequities that are foreseen.

Formal and Informal Economic Analysis

The outline of economic consequences contained in figure 6.2 is likely to intimidate anyone seeking to evaluate a relatively modest wetland project. It may appear to bury a careful decision maker or interested citizen in an impossibly complex formal analysis. That, of course, is not the intent. Figure 6.2 outlines the types of questions that need to be asked and answered, at least informally, about any project. The initial focus should be to determine a full list of the significant changes that will be associated with a wetland management decision. For many small or modest projects, the changes are also likely to be small or modest. This will lead the matrix outline in figure 6.2 to have many cells that can safely be left empty because there are no significant impacts in those categories. Even where one expects some measurable impacts, the size of the project and the funding available for the evaluation will dictate how detailed the economic analysis should be. Economics does not demand any particular level of empirical analysis or quantification; it seeks only to systematically reveal the trade-offs that are unavoidably implicit in wetland management so that decisions can be more fully informed.

At a minimum, this requires that the flow of goods and services from both the natural wetland system and related human activities that are now supported by the status quo be compared with the flow of goods and services that would result from a proposed change in wetland management. Of course, besides this "with" and "without" comparison, one could also include comparisons between other alternatives that are relevant. The point is to have a side-by-side comparison of all the significant changes, both positive and negative, associated with the proposed project as well as those associated with "doing nothing."

At its most informal level, all that is required for this type of trade-off analysis is a listing of the significant changes expected from the management change, no matter what type of value they might impact: environmental, narrowly economic, recreational, cultural, equity, and so on. If people are likely to be concerned about the size and/or type of change resulting from wetland management choices, they should be included in the analysis.

No particular level of quantification of impacts is necessarily required. In its crudest form, the changes can simply be described as accurately and objectively as possible. If it is possible to merely rank the impacts by relative importance (such as high, medium, and low or on a scale of 1 to 10), that may be appropriate for some affected values. Some impacts may be quantifiable in physical terms (such as acres disturbed, volume of peak flows, and change in nutrient load). Finally, some of the impacts may be expressed in dollar value terms either because they represent goods and services traded in commercial markets (such as food and forage, commercial recreation, and electricity) or because economic analysis has generated such dollar values for nonmarketed goods and services (such as recreation, flood control, water quality, and wildlife). Clearly, there is quite an array of possible ways of expressing the positive and negative consequences of any particular wetland management alternatives and systematically evaluating them.

This type of evaluation is "pluralistic" in the sense of allowing a variety of different expression of relevant values. It does not seek to reduce all valuation to a single measure such as dollars. Some degree of such pluralism in the expression of values is almost always necessary in wetland economic analysis for several reasons. First, there is no agreement among environmental analysts that all relevant human values can be reduced to a common measure and then compared to one another. Some relevant human values may simply be incommensurable (Sagoff 1996). Economists, environmental philosophers, and ecological scientists continue to debate this issue. Second, even if all relevant values were comparable in common terms, we are unlikely to know enough about the physical impacts of a particular project, the effect of those physical impacts on relevant human concerns, and the value of the changes. Our knowledge will be incomplete, and, as a result, a comprehensive formal economic valuation simply will not be possible. Rather than ignore the impacts that could not be monetized in the economic valuation, it is far better to list the impacts themselves in whatever metric is available. Finally, for small projects, the amount of economic and other analysis that is appropriate is likely to be quite limited. Even if a comprehensive economic evaluation of all aspects of the project were conceptually feasible, it might not be cost effective. In that setting, a pluralistic expression of values in a largely informal analysis may be exactly what is appropriate.

Examples of Wetland Economic Studies

ESTIMATED NONCOMMERCIAL WETLAND ECONOMIC VALUES

Since markets will not quantify the economic values associated with many noncommercial changes in environmental stocks and flows, the economic analyst has two choices. The first is to complete the quantitative analysis using only market values and then describe in nonmonetary terms the nonmarket economic costs and benefits that were left out of the economic analysis and discuss how those nonmarket values might modify the conclusions. The second approach that can be taken is to estimate at least some of the economic values associated with the nonmarket environmental resources in money terms. Over the past half century, environmental and natural resource economics has developed a variety of tools for estimating nonmarket economic values that are comparable to those economic values reflected in market outcomes. This dollar quantification of environmental values has come to be both required by many federal regulations for large projects and accepted by the American judicial system as useful approximations of real economic benefits and costs (National Oceanic and Atmospheric Administration 1993; U.S. Water Resources Council 1979; Ward and Duffield 1992). Of course, for relatively modest restoration projects, costly efforts to quantify in dollar terms the environmental impacts of the project may not be feasible or appropriate.

Table 6.1 lists some of the typical values estimated by economists for wetland nonmarket economic values as well as the methods typically used for those estimates. When a change in wetland function changes the costs faced by economic actors, those increased or reduced costs can be used to estimate at least part of the change in economic values. For instance, if a change in the management of wetlands increases flooding or reduces water quality, the damage done by increased flooding or the cost of treating the water to return it to an acceptable quality for use could be used to estimate some of the economic values associated with the change. This is the "avoided damage" approach. Somewhat more removed from direct market information is the use of "replacement cost" to estimate damages. This approach estimates what the cost would be to obtain lost or degraded ecosystem services if the least-costly method to mitigate or eliminate the damage was adopted. For instance, if increased flooding were the result, the cost of downstream flood control measures could be used to estimate the cost of the increased flooding. Finally, some of the impacts of wetland function change are expressed in dollar terms using travel cost as well as other be-havioral (such as hedonoic) studies as well as contingent valuation meth-ods discussed earlier.

Estimates of the Economic Value of Global Wetland Functions

In 1997, an international group of ecological economists published an in-fluential paper in the international science journal *Nature* that sought to estimate the economic value of the environmental services provided to hu-mans by natural systems, including wetland and other water-based natural systems (Costanza et al. 1997). Because of the huge size of the estimated economic values, the article received considerable attention in the popular media around the world as well as in other scientific journals.

Using a mix of all the tools discussed previously for estimating wetland economic values, these economists estimated the total value of all ecosys-tem services to be about $33 trillion a year. Wetlands in the form of estu-aries, tidal marshes/mangroves, and swamps and floodplains contributed over a quarter of this total economic value, about $9 trillion a year. Econ-omists have debated the accuracy of various components of these esti-mates. The authors of the study, however, insist that they have made their main point with reasonable accuracy: Natural systems, including wet-lands, produce economically valuable services on a scale that can be ig-nored only at the threat of serious losses to the well-being of the planet's population.

Wetlands and Flood Control in the Great Plains

Some of North America's most extensive wetlands are found on the rela-tively flat prairies of the Great Plains. Over time, many of these wetlands have been drained and converted to cropland. One result has been an increase in flooding as the land lost its capacity to absorb and slowly release natural precipitation. To cope with the flooding, rivers were cleared, dredged, and channelized so that the floodwaters could be moved more quickly away from valuable farmlands and urban areas. This tended to move larger floods downstream. One solution to this increased flooding was to design and build dams to hold back the floodwaters. These dams, in turn, permanently flooded river valleys and the natural systems associated with them under large, slow-moving reservoirs. These dams and reservoirs, in turn, created additional environmental problems.

Ecologists, environmentalists, fiscal conservatives, and economists have raised questions about the practical, biological, and economic rationality of this ongoing string of unintended consequences that has yet to come to an end. Many studies have been conducted of the conflict between the private rationality of draining the wetlands on a farm and the social rationality of the consequences that private action triggers. The landowner acts to enhance the productivity of his or her land by passing on substantial costs to downstream landowners and the general public. Many economic studies have concluded that the most cost-effective flood control consists of restoring wetlands and reducing the channelization and/or damming of rivers (Johnson 1997; Leitch and Hovde 1996; Power and Niemi 1998; Roberts and Leitch 1997). Maintaining coastal wetlands can also reduce the flood damage caused by coastal storms (Farber 1996; Rager et al. 1995).

Use of Wetlands to Protect Water Quality

Wetlands have the capacity to assimilate and immobilize a variety of chemical and biological agents that threaten human and other species health. If this service is not provided by wetlands, either people have to put up with the damage caused by the water pollution or costly water treatment facilities have to be constructed. Many economic studies have compared the use of existing wetlands or the restoration of wetlands to the construction of conventional water treatment facilities. Many of them have concluded that protection or restoration of wetlands is the most cost-effective way to maintain or improve water quality. Anderson (1993) found that using wetlands to treat the discharge from a sugar beet refining operation cost only about one-fifth of what a conventional treatment facility would cost. Gillette (1994) found that wetlands were effective in treating the wastewater from an aluminum processing plant. Andréasson-Gren (1991) found that the use of wetlands to reduce nitrogen sources that were contributing to eutrophication of water bodies in Sweden was "the undoubtedly cheapest alternative." Other studies have found that wetlands are cost-effective ways of removing phosphorous from both municipal water effluent as well as drainage from agricultural lands (Kunkel and Steel 1993; Schaefer et al. 1996). Breux (1992) analyzed the use of Louisiana coastal marshlands for the tertiary treatment of wastewater. Wetlands were also found to be cost effective in the treatment of coal mine discharges (Baker et al. 1991; Skousen et al. 1992).

Conclusion

In popular discussions, economics has often been mischaracterized as focused exclusively on the world of money and commercial business transactions and criticized for ignoring environmental and other noncommercial values. As environmental economists have made headway estimating noncommercial economic values, others have instead criticized economics for even trying to give economic expression to environmental values. They doubt the reliability and/or the appropriateness of estimating environmental economic values.

On the innovative frontiers of all scientific fields, there is controversy. That is also true in environmental and wetland economics when it comes to quantitative estimates of particular environmental values. But such skepticism about estimating the dollar value of noncommercial wetland functions should not be confused with a rejection of economic analysis in the evaluation of wetland management policies. As the examples in this chapter demonstrate, economics offers a far broader range of tools and insights than just dollar valuation of noncommercial economic values. Economics focuses on improving the way we use scarce resources that have alternative uses as we pursue our diverse needs and desires. Decisions about the management of wetlands certainly fall within this realm.

Economics can contribute in all of the following ways to improving wetland management:

1. Specifying the linkages between wetland system functioning and the full range of human objectives.

2. Organizing the information on the full range of costs and benefits associated with wetland management alternatives, expressed in whatever terms are practical.

3. Analyzing the unavoidable trade-offs between the pursuit of various objectives, such as boosting farm incomes, stimulating the local economy, flood control, protecting water quality, and providing wildlife habitat and recreation opportunities.

4. Determining the cost-minimizing path to achieving socially determined wetland management objectives, such as water quality standards or endangered species protection.

5. Understanding the ways in which public policies can encourage uneconomic waste of both environmental and commercial resources, such as public policies that encourage converting all potential lands to agriculture and then committing enormous resources to mitigating the damage that results.

6. Understanding the ways that incentive systems can motivate socially unproductive human action, such as draining or destroying valuable wetlands.

7. Understanding the alternative ways in which human goals can be pursued; a "requirements" or "materials flow" approach to the economy ignores the existence of substitutes, the development of new technologies, and the overall adaptability of people and economies.

8. Critical understanding the advantages of maintaining decentralized decision-making systems that are efficient and responsible and that allow the broadest range of innovation and choice; such systems are not always appropriate, but economics can critically evaluate alternative institutional arrangements for economic decision making.

9. Estimating the value of noncommercial wetland goods and services so that they can be compared with the market values that wetland conversion can support.

Given the contribution economics can make to rational management of our wetland resources, it would be a mistake to reject its use on the basis of an inappropriately narrow understanding of the economist's approach to natural resource and environmental policy.

REFERENCES

Abdalla, C. W. 1990. Measuring economic losses from ground water contamination: An investigation of household avoidance costs. *Water Resources Bulletin* 26(3):451–63.

Anderson, P. 1993. Constructed wetlands are a sweet deal. *Water, Environment and Technology* 5(7):56–59.

Andréasson-Gren, I.-M. 1991. Costs for nitrogen source reduction in a eutrophicated bay in Sweden. In *Linking the Natural Environment and the Economy*. Edited by C. Folke and T. Käberger. Dordrecht: Kluwer, 173–88.

Baker, K. A., M. S. Fennessey, and W. J. Mitsch. 1991. Designing wetlands for controlling coal mine drainage: An ecologic-economic modelling approach. *Ecological Economics* 3:1–24.

Breux, A. M. 1992. The use of hydrologically altered wetlands to treat wastewater in coastal Louisiana. Ph.D. diss., Louisiana State University.

Costanza, R., R. d'Arge, R. de Groot, S. Farber, M. Grasso, B. Hannon, K. Limburg, S. Naeem, R. V. O'Neill, J. Paruelo, R. G. Raskin, P. Sutton, and M. van den Belt. 1997. The value of the world's ecosystem services and natural capital. *Nature* 387:253–60.

Courant, P. N. 1994. How would you know a good economic development policy if you tripped over one: Hint: Don't just count jobs. *National Tax Journal* 47(4):863–81.

Duffield, J., C. Neher, and T. Brown. 1992. Recreation benefits of instream flow: Application to Montana's Big Hole and Bitterroot Rivers. *Water Resources Research* 28(9):2169–81.

Farber, S. 1996. The economic welfare loss of projected Louisiana wetlands disintegration. *Contemporary Economic Policy* 14(1):92–106.

Gillette, B. 1994. Constructed wetlands for industrial wastewater. *Biocycle* 35(11):80–83.

Greenley, D. A., R. G. Walsh, and R. A. Young. 1981. Option value: Empirical evidence from a case study of recreation and water quality. *Quarterly Journal of Economics* 96(4):657–73.

Johnson, R. R. 1997. *The Vermillion River: Managing the watershed to reduce flooding.* Vermillion, S.D.: Clay County Conservation District.

Kunkel, J. R., and T. D. Steele. 1993. Impacts of a natural wetland on total-phosphorus loads downstream from a wastewater treatment plant. In *Basin Planning and Management: Water Quantity and Quality Information 73.* Edited by D. K. Mueller. Fort Collins: Colorado State University, Water Resources Research Institute, 57–64.

Leitch, J. A., and B. Hovde. 1996. Empirical valuation of prairie potholes: Five case studies. *Great Plains Research* 6(1):25–39.

Leschine, T. M., K. F. Wellman, and T. H. Green. 1997. The economic value of wetlands: Wetlands' role in flood protection in western Washington. School of Marine Affairs Working Paper, University of Washington, Seattle.

National Oceanic and Atmospheric Administration. 1993. Natural resource damage assessments under the Oil Pollution Act of 1990. Report of the NOAA Panel on Contingent Valuation. *Federal Register* 58:4601–14.

National Research Council, Committee on Restoration of Aquatic Systems. 1992. *Restoration of aquatic ecosystems: Science, technology, and public policy.* Washington, D.C.: National Academy Press.

Power, T. M., and E. Niemi. 1998. An economic evaluation of flood control alternatives in the Vermillion River Basin, South Dakota. *Great Plains Natural Resources Journal* 3(1):3–71.

Rager, K. A., C. B. Clifton, and L. T. Johnson. 1995. San Diego County wetlands: History, inventory, ecology, and economic valuation with specific reference to agricultural nonpoint source pollution. Publication UCSGEP-SD- 95-1. University of California Sea Grant Extension Program, San Diego.

Roberts, L. A., and J. A. Leitch. 1997. Economic valuation of some wetland outputs of Mud Lake, Minnesota-South Dakota. Agricultural Economics Report 1. North Dakota State University, North Dakota Agricultural Experiment Station, Department of Agricultural Economics, October.

Sagoff, M. 1996. On the value of endangered and other species. *Environmental Management* 20(6):897–911.

Schaefer, K., E. Snell, and D. Hayman. 1996. Valuing wetland nutrient buffers in the Erasmosa River watershed. In *Developpement durable et rationnel des ressources hydriques: Compte rendu de la 49e Conférence annuelle de l'Association canadienne des ressources hydriques,* vol. 2. Edited by C. E. Deslisle and M. S. Bouchard. Collection Environnement de l'Université de Montréal. Montreal: L'Université de Montréal, 62–638.

Skousen, J. G., T. T. Phipps, and J. Fletcher. 1992. Acid mine drainage treatment alternatives. Land Reclamation: Advances in Research and Technology ASEA Publication 14-92. St. Joseph, Mich.: American Society of Agricultural Engineers, St. Joseph, Michigan, 297–303.

Thibodeau, F. R., and B. D. Ostro. 1981. An economic analysis of wetland preservation. *Journal of Environmental Management* 12:19–30.

U.S. Water Resources Council. 1979. Principles, standards and procedures for water and related land resource planning. *Federal Register,* no. 242(December 14):72978–72990.

———. 1983. Economic and environmental principles and guidelines for water and related land resources implementation studies. March 10.

Walsh, R. G., D. M. Johnson, and J. R. McKean. 1988. Review of outdoor recreation economic demand studies with nonmarket benefit estimates, 1968–1988. Technical Report 54. Colorado State University, Colorado Water Resources Research Institute, Fort Collins.

Ward, K. M., and J. W. Duffield. 1992. *Natural resource damages: Law and economics.* Somerset, N.J.: Wiley.

SUGGESTED READINGS

Goodstein, Eban S. *Economics and the environment.* 3rd ed. Wiley, 2002.

Kahn, James R. 1995. Square pegs and round holes: Can the economic paradigm be used to value the wilderness? *Growth and Change* 26(fall):591–610.

Power, Thomas Michael. 1996. *Environmental protection and economic well-being: The economic pursuit of quality.* Armonk, N.Y.: M. E. Sharpe, 1996.

———. 2001. The contribution of economics to ecosystem preservation: Far beyond monetary valuation. In *Managing Human-Dominated Ecosystems*. Monographs in Systematic Botany from the Missouri Botanical Gardens, vol. 84. Edited by Victoria C. Hollowell. St. Louis: Missouri Botanical Gardens Press.

Scodari, Paul F. 1997. *Measuring the benefits of federal wetland programs.* Washington, D.C.: Environmental Law Institute.

Concluding Thoughts

THE CHALLENGE
OF PRESERVING
WETLANDS

Sharon L. Spray
and
Karen L. McGlothlin

We may have progressed significantly from the years in which historian William Myer described the wide-scale state-sponsored elimination of valuable wetlands across the country. We more fully understand the benefits that these valuable ecosystems contribute to water quality, flood control, wildlife habitat, and pollution mitigation than we did a half century ago, but throughout the United States and the rest of the world, wetlands continue to disappear at alarming and unsustainable rates.

Wetlands are public goods that provide widely dispersed benefits, and because the benefits of wetlands are often not well understood by citizens or policymakers, protection of these ecosystems continues to be extraordinarily challenging. As Thomas Power points out in this volume, when making a decision about conversion of a wetland to another use, citizens must often weigh the *potential* long-term consequences of wetland destruction with the often immediate benefits associated with proposed transformations. For instance, land transformation projects that bring new housing to growing communities and new industrial parks that bring jobs to lagging economic regions have concentrated, tangible net returns for citizens. When citizens have extremely low levels of knowledge about wetland functions, it can be politically difficult for decision makers to advocate protection.

Bardes and Oldendick (2000) indicate that Americans are supportive of environmental protection but that support drops off considerably if they believe that environmental protection will impact their standards of living. Since the dispersed benefits of wetlands are often juxtaposed against the concentrated and immediate benefits of other land uses, how vigorously the nation should protect wetlands is as much a scientific question as it is a political and economic one.

Wetlands protection in the United States is complicated by federal and state regulatory conflicts and constitutional protections accorded property owners. In the United States, approximately 75% of wetlands are privately owned. Permitting rule changes under the second Bush administration have favored energy companies, commercial and residential developers, and small-property owners who have long fought to ease restrictions on the use of wetland regions. New permitting rules that went into effect in January 2002 will make it easier for developers to obtain permits to fill wetlands in floodplains and transform wetlands for commercial and residential use. Under new permitting rules, developers may now obtain permits to destroy a half acre of wetlands per lot of commercial property rather than limiting wetland destruction to a half acre per commercial project (League of Conservation Voters 2003). This change could exponentially increase the total acreage of wetlands destroyed in large-scale projects in which multiple lots are combined for use in a single project (such as several lots purchased as individual units but combined for development of a shopping mall or an airport).

Several recent court cases have also measurably diminished the scope of national legislation governing wetlands and will likely lead to increased losses of wetlands throughout the country unless Congress strengthens existing laws with more specific goals and defined statutory language. One such case that reshaped wetlands regulations in recent years is the January 2001 U.S. Supreme Court decision in the case of *Solid Waste Agency of Northern Cook County v. United States Corps of Engineers* (*SWANCC v. USACOE*). In this case, the Court stated that the Corps of Engineers (hereinafter Corps) exceeded its jurisdictional authority under Section 404 of the Clean Water Act (CWA) to regulate *isolated* wetland habitats and thereby widely curtailed protection of wetlands across the nation.

As discussed by Mary Hague in this volume, the CWA provides the broad federal authority for protection of wetlands, but the specific regulations for wetland protections evolve from the Corps's interpretation of the CWA and the regulatory guidelines it has implemented to meet the statute's goals. The statutory language of the CWA refers to protection of "navigable waters" with broad constitutional authority found in the commerce clause of the U.S. Constitution. Protection of navigable waters was extended to include wetlands adjacent to our rivers and tributaries as a consideration of the value that wetlands play in the environmental health and stability of those waters.

For decades, the Corps has broadly interpreted the 1972 Water Pollution Control Amendments to the CWA to include protection powers for *all* the nation's water resources and aquatic ecosystems, including the protection of isolated lakes and wetland areas that may or may not be adjacent to or near navigable waters. Although the CWA does not specifically state that isolated wetlands must be protected, the Corps's interpretation of the CWA is firmly based on our scientific understanding of the interconnections between water ecosystems, whether or not a water system, such as a meadow bog or a summer marshland, is used for navigable purposes. Until this decision, the Corps interpreted the CWA to include the following:

Waters such as intrastate lakes, rivers, streams (including intermittent streams), mudflats, sandflats, wetlands, sloughs, prairie potholes, wet meadows, playa lakes, or natural ponds, the use, degradation, or destruction of

which could affect interstate or foreign commerce. (33 CFR 328.3[a][3]
[1999], quoted in Kunsler 2003, 3)

In 1986, the Corps further clarified its jurisdiction over wetlands by adopt-
ing a "migratory bird rule." This rule emerged to assist the Corps in meeting U.S. obligations under several multilateral agreements signed in the
past quarter century to protect migratory birds. As author Joel Snodgrass
in this volume indicates, there are many species of animals and migratory
birds that rely solely on wetlands for reproduction and foraging. Wetlands
provide many bird species with protection from predators and highly concentrated food sources necessary for survival. One of the most important
of these treaties is the 1979 Convention on the Conservation of Migratory
Species of Wild Animals (CMS), also know as the Bonn Convention,
which facilitates the preservation of endangered animals, including birds,
whose migratory patterns cross national boundaries and are dependent on
more than one habitat for survival. The migratory bird rule has also been
instrumental for implementation of U.S. treaty obligations under the 1986
North American Waterfowl Management Plan designed to protect the diversity of ducks whose numbers reached record lows during the 1980s.

    In *SWANCC v. USACOE*, the Corps denied a permit to build on land
not clearly defined as a wetland but used by migratory birds typically dependent on wetland ecosystems. The Supreme Court ruled that the
Corps's interpretation of the CWA using the migratory bird rule was
overly broad. The Court held that Congress never intended Section 404 to
apply to isolated waters based merely on the presence of migratory birds.

    The Court's ruling in *SWANCC v. USACOE* could limit federal author-
ity to only about 20% of the nation's current wetlands. "Under [a strict
interpretation], wetlands regulated under the Clean Water Act would
primarily include river fringing wetlands for larger rivers and streams,
lake fringing wetlands for larger lakes, and coastal and estuarine fringing
wetlands" but would not cover important wetland ecosystems such as
"prairie potholes, wet meadows, river fringing wetlands along small, non-
navigable rivers and streams, many forested wetlands, vernal pools, seeps

and springs, flats, bogs and large amounts of tundra in Alaska" (Kunsler 2003, 6). A broader interpretation of the ruling could raise the acreage of wetlands falling under the jurisdiction of the CWA significantly, but there is little doubt without additional legislation, as a result of this ruling, that fewer wetlands now will be protected under the CWA than were only a few years ago, and none of this is based on a change in scientific findings.

The Court also potentially narrowed the federal government's ability to regulate the use of isolated wetlands not adjacent to navigable waters when it stated the following: "Permitting respondents to claim federal jurisdiction over ponds and mudflats falling within the 'Migratory Bird Rule' would result in significant infringement of the State's traditional primary power over land and water use" (Kunsler 2003, 4). This statement could signal the Court's greater willingness to ease federal environmental regulations and let states make more environmental decisions. Many environmentalists fear that turning over the regulation and conservation of wetlands to state governments would be disastrous.

Developers, oil and gas interests, and agribusinesses have fought vigorously in recent years to relax wetland laws and shift federal authority back to the states (see Hinkel 1999). They often suggest that state governments are better suited to manage environmental problems within state boundaries. After all, states vary considerably in their density of population, their agricultural crops, and even their climate. Special interests thereby conclude environmental issues would be best addressed by those government officials familiar with these idiosyncrasies. Yet environmentalists and political scientists point out that local control may not yield better outcomes. Walter Rosenbaum (1998) suggests that "by encouraging legislators to view environmental proposals first through the lens of local interests, it often weakens sensitivity to national needs and interests. At worst it drives legislators to judge the merits of environmental policy almost solely by their impact on frequently small and atypical constituencies" (70).

The basis for existing federal laws that govern the use of wetlands emerges from the recognition that without federal regulation there would continue to be wide variance in state regulation of waterways and water resources between states; and because the conservation of wetlands is a public good and destruction of wetlands will have far-reaching effects beyond each state's borders, federal regulation is necessary. In many cases, states have enacted state laws that have stricter standards than that used with the permitting process under the CWA, and many have adopted state watershed planning projects that have greatly contributed to improved water quality and ecosystem protection. But there are also many states that have no staffed or funded programs to protect isolated wetlands or freshwater wetland systems that fall outside the scope of existing federal laws. Many scientists and environmentalists worry that delegating more authority to the states for management of wetlands will likely result in less wetland protection rather than more.

The first Bush administration's establishment of the 1989 federal "no net loss of wetlands" policy (see Hague, this volume) was designed to address the growing hostility of private landowners and other special interests that were unhappy with federal restrictions placed on private and public land use. The no-net-loss policy added **flexibility mechanisms** into the permit process allowing the creation, restoration, or enhancement of wetlands in exchange for wetland losses in some areas. The policy is explained by the National Research Council (2001):

> When there is a proposal to discharge dredged or fill material into a wetland, the CWA expects that the Corps, in cooperation with other agencies, will consider the public-interest consequences of issuing a permit. In practical terms, implementation of Section 404 and related programs has followed a general policy that the deliberate discharge of materials must be avoided where possible and minimized when unavoidable. Then if a permit is issued, and wetland functions are compromised, some kind of **compensatory mitigation** may be required to replace the loss of wetland's functions in the watershed. (12)

While the policy is laudable for its attempt to balance competing social and economic needs, with the maintenance of current levels of national wetlands, it has been highly controversial in the scientific and policy communities that continue to debate whether such an approach to wetland conservation is actually sound. After nearly a decade and a half, the outcome of the no-net-loss policy path is dismal. Loss rates continue at nearly 60,000 acres annually. Natural and social scientists are raising a number of questions about whether these flexibility mechanisms are truly suitable for stabilization or preservation of existing levels of national wetlands.

Compensatory mitigation projects have several forms. As a condition for a permit, the recipient of the permit may agree to offset wetland drainage through a **wetland restoration** project in which wetlands previously degraded or altered by human activity are restored. They can engage in a **wetland construction** project in which a wetland area is created where none previously existed with the goal of pollution removal and wastewater runoff mitigation. **Wetland preservation** projects are also sometimes undertaken as compensatory projects. In such projects, the recipient of the permit protects a fully functioning wetland from possible future threats, or the beneficiary may be asked to engage in a **wetland enhancement** project in which one or more wetland functions are enhanced in an existing wetlands area. In some cases, permit recipients use third-party mitigation strategies. "In third-party mitigation (i.e., commercial mitigation bank, in-lieu fee program, cash donation, or revolving fund program), another party accepts a payment from the permittee and assumes the permittee's mitigation obligation" (National Research Council 2001, 2). Yet it is clear that there are continuing problems with all these compensatory programs.

As John Callaway points out in chapter 3, the creation of new wetlands or the restoration of degraded wetlands is not as easily accomplished as once believed. While compensatory mitigation was hailed as a brilliant alternative to the loss of natural wetlands to development in the early 1980s, it quickly became apparent that, in many cases, the wetlands ecosystem is too complex to accurately duplicate (Roberts 1993). The literature on compensatory mitigation is replete with examples of created or restored wetlands that failed to function in a manner similar to the natural wetlands that they were constructed to replace. A greater problem with flexibility mechanisms may be the low level of accountability associated with these projects. The Washington State Department of Ecology estimates that only 13% of the man-made wetlands in their state are fully successful. To make matters worse, several studies have found that a large percentage of mitigation projects, in some cases up to 80%, have never been started at all (Froelich 2003).

This volume was never designed to encourage specific policy approaches for wetlands preservation, but we do want readers to recognize that current policies remain inadequate. Wetlands function and sustain species differently than any other type of ecosystem, and the consequences of their continued destruction will affect the environmental quality for current and future generations in measurably problematic ways. As the authors in this volume point out, we still lack the knowledge to effectively recreate wetlands, making their conservation extremely important if we wish to preserve the benefits that we accrue from wetlands before all or most of them are lost forever.

From both an aesthetic and an ecological standpoint, the world would be far less rich without wetlands. Environmental engineer Alice Outwater (1996) writes, "In a wetland, the food web is dense and the niches are varied. Frogs twang in the evening, warning of a raccoon wading out to dig up grubs and insects. Herons stalk the frogs, and migrating ducks settle out of the sky to rest and feast before traveling on. Meadowlarks and magpies alight upon the stumps, and muskrats, voles, and otters make their homes along the shore. Sometimes a moose or a deer wades into the water to eat the greens along the shore, while minnows hide among the stalks" (25).

Effective environmental policy requires asking, What kind of world do we want to live in? Through much of our history, we have failed to acknowledge the value inherent in wetland ecosystems and in the process have compromised the quality of our water resources and the services provided by them. As Outwater (1996) explains, "By dredging, by damming, by channeling, by tampering with (and in some cases eliminating) the ecological niches where water cleans itself, we have simplified the pathways that water takes through the American land-scape; and we have ended up with dirty water" (xii). We must learn from past policies and create future policy that leads us to better results than those of our past and present that may be well-intentioned but prob-lematic in result.

As we experience greater flooding, degraded water quality, diminished biodiversity, and reduced food stocks, policymakers will eventually be forced to recognize that we cannot ignore or undervalue either the direct or the indirect costs of wetlands conservation as we calculate the costs and benefits of policy choices. We hope that we will adopt new strategies for greater conservation long before we have lost so many of the nation's wet-lands that the environmental consequences are irreparable.

There are practical steps that can be taken to preserve our nation's wet-lands, including greater enforcement of permit obligations, modification of the CWA to fully protect isolated wetlands, and implementation of mandatory state watershed planning requirements supported by federal funding. We must also continue to support scientific research on wetlands, including inventories and long-term studies of wetlands enhancement and restoration projects. But what may be needed most is greater public un-derstanding of the value of wetland resources to the nation's long-term en-vironmental health and greater political vigilance. We hope that this vol-ume has increased readers general understanding of the environmental, economic, and social values of wetland ecosystems and that they will use this knowledge to design and promote policies to sustain these valuable environmental treasures.

REFERENCES

Bardes, Barbara A., and Robert W. Oldendick. 2000. *Public opinion: Measuring the American mind.* Belmont, Calif.: Wadsworth.

Froelich, Adrienne. 2003. Army Corps: Retreating or issuing a new assault on wetlands?" *Bioscience* 53(2):130.

Hinkel, Maureen Kuwano. 1999. In our hands. *Forum for Applied Research and Public Policy* 14, no. 3(fall):66–72.

Kunsler, Jon. 2003. The SWANCC decision and state regulation of wetlands. Association of State Wetland Managers, Inc. www.aswm.org.

League of Conservation Voters. 2003. www.lcv.org.

National Research Council. 2001. *Compensating for wetland losses under the Clean Water Act.* Washington, D.C.: National Academy Press.

Outwater, Alice. 1996. *Water: A natural history.* New York: Basic.

Roberts, Leslie. 1993. Wetlands trading is a losing game, say ecologists. *Science* 260(5116):1890–93.

Rosenbaum, Walter A. 1998. *Environmental politics and policy.* 4th ed. Washington, D.C.: Congressional Quarterly Press.

SUGGESTED READINGS

Dahl, T. R. 2000. *Status and trends of wetlands in the conterminous United States 1986 to 1997.* Washington, D.C.: U.S. Department of the Interior, U.S. Fish and Wildlife Service.

Reynolds, John E., and Alex Regalado. 2002. The effects of wetlands and other factors on rural land values. *Appraisal Journal* 70, no. 2(April):182–90.

U.S. Environmental Protection Agency. 2003. www.epa.gov/owow/wetlands.

Appendix

Redox reactions are represented by the following reduction equation:

$$a\text{OX} + b\text{H}^+ + n\text{e}^- \rightarrow c\text{RED} \qquad (1)$$

where OX is the oxidized component or electron acceptor; RED is the reduced component or electron donor; a, b, and c are the stoichiometric coefficients for the species involved in the reaction; and n is the number of electrons involved in the reaction. Like all chemical reactions, the reduction reaction can be described quantitatively through the change in Gibbs free energy (ΔG),

$$\Delta G = \Delta G^\circ + RT \ln \frac{(\text{Red})^c}{(\text{Ox})^a (\text{H}^+)^b} \qquad (2)$$

where ΔG° is the standard free energy change, R is the gas constant, T is absolute temperature in degrees Kelvin (K). The Nernst equation converts the free energy of the reduction reaction to electrochemical potential using the relationship $\Delta G = -nEF$,

$$Eh = E^\circ - \frac{RT}{nF} \ln \frac{(\text{Red})^c}{(\text{Ox})^a (\text{H}^+)^b} \qquad (3)$$

where Eh is the electrode potential (in millivolts), E° is the standard half-cell potential, F is the Faraday constant, n is the number of electrons exchanged in the half-cell reaction, b is the number of protons exchanged, and the activities of the various oxidized and reduced components are shown in parentheses. Substituting values of 8.31 J $Kmol^{-1}$ for R (gas constant), 9.65×10^4 cal mol^{-1} for F (Faraday constant), and 298 K for T (absolute temperature), using the relationship $\ln(x) = 2.303 \log(x)$, and converting H^+ to pH, equation 3 simplifies to

$$Eh = E° - \frac{59}{n} \log \frac{(Red)}{(Ox)} - 59\left(\frac{b}{n}\right)pH \qquad (4)$$

The mathematical relationships in equation 4 demonstrates that the redox potential (Eh) increases with increasing activity of the oxidized component, decreases with increasing activity of the reduced component, and increases with an increase in hydrogen ion activity (or a decrease in pH). Simply put, strongly oxidized systems have a positive Eh (e.g., +400 mV) and strongly reduced systems have a negative Eh (e.g., −100 mV).

The electrons used in the reduction equation 1 must be supplied by an accompanying oxidation reaction. In soils, organic matter (CH_2O) is the primary source of electrons. Therefore, a complete redox reaction balances the reduction reaction with an appropriate oxidation reaction. Equations 5 to 7 illustrate this approach by coupling the reduction of $Fe(OH)_3$ with the oxidation of CH_2O:

$$4Fe(OH)_3 + 12H^+ + 4e^- \rightarrow 4Fe^{2+} + 12H_2O \text{ (reduction)} \qquad (5)$$
$$\underline{CH_2O + H_2O \rightarrow CO_2 + 4H^+ + 4e^- \text{ (oxidation)}} \qquad (6)$$
$$4Fe(OH)_3 + CH_2O + 8H^+ \rightarrow 4Fe^{2+} + CO_2 + 11H_2O \qquad (7)$$

Redox potential can also be quantified as pe, $-\log(e^-)$, in the same way pH is defined as the $-\log(H^+)$; however, Eh is the more common expression of soil redox potential since it is readily determined with a platinum (Pt) electrode. The Pt electrode is the most suitable, as it readily transfers electrons either to or from the medium but does not react with them, allowing an accurate measure of the electron flow. There are several methods of Pt-electrode construction, but they all follow the same basic design of fixing Pt wire directly to copper wire or brass rod (Faulkner and

Richardson 1989). Reduced soils transfer electrons to the Pt electrode, while oxidized soils tend to take electrons from the electrode. For actual redox potential measurements, electron flow is prevented, and the potential between the half cell composed of the platinum in contact with the substrate and the known potential of the reference electrode half cell is measured with a suitable meter that responds to electromotive force or potential. Redox potential measurements are made using a portable pH/ millivolt (mV) meter and a saturated calomel or silver/silver-chloride reference electrode.

R E F E R E N C E

Faulkner, S. P., and C. J. Richardson. 1989. Physical and chemical characteristics of freshwater wetland soils. In *Constructed wetlands for wastewater treatment*. Edited by D. Hammer. Chelsea, Mich.: Lewis, 41–71.

Glossary

adaptive management—a wetlands management approach in which monitoring information is used to iteratively revise management decisions and goals.

adsorbed—a condition where ions are attached to the surface of soil particles.

aerobic respiration—the breakdown and release of stored energy from organic matter using oxygen as the terminal electron acceptor.

agenda setting—one of the first steps of the public policy process; this is an outgrowth of problem identification that results in political institutions considering action to address a problem. Following agenda setting, the steps of formulating, legitimating, implementing, and evaluating constitute the remainder of the policymaking process. Formulating policy consists of drafting legislation on the basis of the review and selection of policy options. Policy is legitimated when the appropriate political institutions endorse, support, or adopt the policy. Implementing refers to the actions taken by bureaucracies, state and local governments, as well as the private sector to comply with the policy's provisions. Finally, evaluating refers to analysis of the policy in order to reform or terminate its provisions.

age structure—the distribution of individuals among age classes.

anaerobic conditions—a condition where no oxygen is present.

anaerobic respiration—the breakdown and release of stored energy from organic matter in the absence of oxygen, requiring the use of alternate electron acceptors.

anaerobiosis—the absence of oxygen.

anoxic—condition in which free oxygen, gaseous or dissolved, is either deficient or absent (*see also* anaerobic conditions).

assimilatory sulfate reduction—a metabolic pathway where sulfide produced by microbial sulfate reduction is incorporated into the cellular structure of the microorganisms.

biological control—the control of pest organisms by the introduction of natural predatory organisms or their products.

biological oxygen demand—the amount of oxygen required by aerobic microorganisms to decompose the organic matter in water.

biotic environment—the collection of organisms of the same or different species frequently interacting with an individual, population, or species in a defined area or habitat.

bog—a type of precipitation-dominated wetland that is characterized by the presence of peat deposits, acidic water, and a ground cover of sphagnum moss.

brackish—a term used to refer to water that is a mixture of freshwater and salt water.

Clean Water Act—The Federal Water Pollution Control Act of 1972 and its 1977 amendments are known as the Clean Water Act. The policy sets goals for water quality and creates a permitting process for discharge of water pollutants dealing with surface waters of the United States. The act is implemented by state governments and the U.S. Environmental Protection Agency. Its Section 404 includes provisions interpreted by the courts as allowing for the regulation of wetlands.

community—a group of organisms occurring in the same area or habitat and representing a number of different species that interact on a frequent basis.

compensatory mitigation—the restoration, creation, enhancement, and/or preservation of other wetlands as compensation for impacts to natural wetlands.

concentration gradient—a pathway between areas of high and low concentration.

concentrations—redoximorphic features resulting from the accumulation of oxidized iron and manganese.

constructed wetlands—wetlands built by humans primarily for treating polluted water.

contingent valuation—a method to determine the economic value of environmental resources and services by questioning a scientific sample of people. People are asked about their willingness to sacrifice if that sacrifice would assure (was contingent on) an improvement in an environmental quality.

Council on Competitiveness—created in 1990 by President George H. W. Bush, the council, under Vice President Dan Quayle's leadership, considered industry claims that compliance with particular regulations imposed too high a cost. The council was terminated by President William J. Clinton in 1993.

demographic rates—the numbers of births and deaths in a population, particularly as they relate to the ages of the individuals in the population.

depletions—redoximorphic features resulting from the reduction and subsequent removal of iron and manganese from zones within the soil.

diffusion—the movement of a substance from a more concentrated to a less concentrated area.

dissimilatory sulfate reduction—a metabolic pathway where sulfate is reduced to sulfide during anaerobic respiration by obligate anaerobic bacteria.

diversity—the property of a community that considers the number of species present and the evenness of the distribution of individuals among species. Communities with few species, with one species accounting for the majority of individuals, would have low diversity.

economic equity—a term used in reference to issues of economic fairness that result from variations in the distribution of costs and benefits among groups of special public concern, such as low-income households, family farm operators, Native Americans, and future generations.

economic impact—usually the impact on the local economy as measured in terms of changes in jobs, payroll, tax payments, or the dollar volume of business.

economic value—a measure of the capacity of a scarce resource, good, or service to satisfy human needs and desires, measured in terms of what people are willing to sacrifice to obtain it or what they would demand in compensation before they willingly gave it up.

economics—the study of how scarce resources are managed to satisfy people's needs and desires.

ecosystem—a community of organisms along with their physical environment in a defined area.

ecosystem function—the interactions of the parts of an ecosystem, including the exchange of water, nutrients, and energy among the living and nonliving parts of an ecosystem.

ecosystem services—the conditions and processes through which natural ecosystems and the species that constitute them sustain and fulfill human life.

Eh—electrical energy measurement of a redox reaction usually expressed in millivolts (mV).

electron acceptors—elements and compounds that gain electrons during a chemical reaction.

elites—the leaders in society and politics, usually associated with privileges of wealth, education, or technology.

emigrant—a migrant organism that leaves one area to settle in another.

Endangered Species Act—the 1973 federal act that identifies particular species for study or specific protections, depending on categorization as "threatened" or "endangered," to prevent the extinction of that species.

environmental disservices—the many ways in which features of the environment impose some expense or hardship or inconvenience.

environmental history—the study of the relation of human life to its biophysical surroundings throughout the past and up to the present.

Environmental Protection Agency—A federal agency created in 1970 by President Richard Nixon to oversee the coordination and implementation of environmental laws, specifically pollution prevention and remediation policies.

environmental services—the whole range of benefits that features of the environment may provide.

estuarine wetlands—wetlands found along the coasts of oceans and influenced by tides; the salinities of estuarine wetlands fluctuate and may vary from near zero (freshwater) to that of ocean water.

eutrophication—the nutrient enrichment of an area that often changes ecosystem structure or function and leads to decreased water quality. Cultural eutrophication is sometimes used to connote human-induced nutrient enrichment.

eutrophication gradient—an area where nutrient enrichment decreases with increasing distance from the source of the nutrients.

evapotranspiration—water loss to the atmosphere from soil and vegetation.

external impacts (externalities)—changes in economic values that are not coordinated by markets and prices and that, in that sense, are "external" to market transactions. Increased flooding, decreased water quality, loss of wildlife habitat, and loss of noncommercial recreation opportunities due to the draining of wetlands are examples.

facultative aerobes—a class of microorganisms that prefer to use oxygen in aerobic respiration but can use other electron acceptors during anaerobic respiration.

Farm Bills—the major agricultural subsidies and regulatory policies of recent congressional sessions are referred to as that year's Farm Bill; for example, the 1996 Federal Agricultural Improvement and Reform (FAIR) Act is also know as the 1996 Farm Bill.

fauna—the animal species occupying a particular area.

fecundity—the potential number of gametes or propagules produced by an organism.

federal agencies—units of the federal government created to implement the policies and functions of the government.

federalism—a division of powers between levels of government, such as that established by the U.S. Constitution between the national government and the states.

fen—types of wetlands that are similar to bogs in that they accumulate peat but different in that they receive water inputs from groundwater (as opposed to precipitation). The water found in fens is typically less acidic and has higher nutrient levels than water in bogs.

fermentation—a metabolic process where organic molecules serve as both electron donors and electron acceptors, resulting in the incomplete oxidation of the organic matter.

field indicators—soil colors and features resulting from hydromorphic processes that are used to identify hydric soils.

Fifth Amendment—adopted in 1791, the Fifth Amendment to the U.S. Constitution includes "protections of citizens before the law," such as protection from self-incrimination and requirements for due process of law. It also states that private property shall not be taken "for public use, without just compensation."

flexibility mechanisms—legal variations in policy implementation designed to provide increased flexibility for parties to meet legal guidelines.

flora—the plant species occupying a given area.

free energy—a thermodynamic term that indicates the amount of energy available for a system to do useful work at constant temperature and pressure.

functional equivalency—a concept that is used to describe a desire to restore all wetland functions at a mitigated site to a state in which they are indistinguishable from the functions performed at a reference site.

groundwater-dominated wetland—a type of wetland that receives its water supplies from groundwater inputs rather than precipitation sources.

habitat corridor—a relatively narrow area that connects isolated patches of habitat and allows for the safe travel of individuals between these patches.

habitat type—a group of plant species occurring to gather with similar structure; often operationally defined by the investigator.

hedonic valuation—a method to determine the economic value of environmental qualities by studying related market decisions. For example, studying how housing prices vary depending on the levels of air pollution, crime, or noise can provide evidence as to what people are willing to sacrifice to obtain access to clear air, lower risk of crime, or less noisy living environments.

hydric soils—soils formed under conditions of saturation, flooding, or ponding long enough during the growing season to develop anaerobic conditions in the upper part.

hydrogeomorphology—a term used when climate, basin geomorphology, and hydrology are considered as a single entity in reference to a particular wetland.

hydrology—a term used to describe the water depth, frequency of flooding, and the duration of flooding needed to maintain a particular wetland.

hydromorphic processes—soil-forming processes dominated by excess water and anaerobic conditions.

hydroperiod—the annual water-level fluctuations that occur in wetlands; often used to refer to the amount of time surface water is present during the annual hydrological cycle.

hydrophytes—plants adapted to flooded or saturated soil conditions. These plants are capable of pumping oxygen to their roots, where it detoxifies compounds that build up in soils with little or no oxygen present.

hydrophytic vegetation—plants typically adapted for life in saturated soil condition (*see also* hydrophytes).

hypoxia—an oxygen-deficient (less than 2 milligrams per liter) condition in coastal waters resulting from the high oxygen demand associated with the decomposition of increased productivity in response to eutrophication of aquatic ecosystems.

immigrant—a migrant organism that becomes established in an area where it was not previously found.

individualism—the emphasis on the individual as a private person who should be free to act on his or her will.

interest groups—organizations of people with a shared interest or goal for the purpose of influencing political decisions. Greenpeace and the National Rifle Association are examples of interest groups.

karst geology—A distinctive topography formed primarily by dissolution of limestone and gypsum. Typical features are caves, underground streams, and surface streams that end abruptly in sinks. The type locality is in the Dinaric Alps of northwestern Yugoslavia and northeastern Italy.

labile—readily changed by chemical or physical processes.

liberalism—the belief that individual rights and freedoms should be protected by the government while it ensures equal treatment of all. Liberals object to restrictions on civil rights and support government services; in the United States, liberals are often associated with the Democratic Party.

management plan—a plan for the management of a particular species or natural resource that identifies specific goals and actions to be taken in order to meet those goals; often management plans are in document form and are agreed on by scientists, managers, policymakers, and private interests.

market models—models of analysis based on economic characteristics, such as supply and demand and rational behavior.

marsh—areas that are classified as wetlands because the soil is either periodically or continually inundated with water and they have a characteristic flora that is comprised of emergent vegetation species that are adapted to life in hydric soils.

mass transfer—the movement of mass from one storage unit to another.

mesocosm—an enclosed system used in the field to serve as small replicates of natural systems.

metapopulation—a group of populations occurring in different habitat patches with occasional exchange of individuals among populations.

methanogenesis—the metabolic pathway where methanogens use carbon dioxide or organic compounds as terminal electron acceptors in anaerobic respiration, producing methane.

methanogens—a specialized group of obligately anaerobic bacteria that carry out methanogenesis.

microorganisms—organisms of microscopic size that include the bacteria, virus, fungi, microalgae, and protozoan groups.

minerotrophic—receiving minerals and nutrients from groundwater or surface water in addition to precipitation resulting in a relatively nutrient-rich wetland.

mitigation—avoiding, minimizing, rectifying, reducing, or compensating for resource losses. Wetlands mitigation is a three-step process to reduce the loss of wetland acreage and function that includes avoiding, minimizing, and compensating for impacts to wetlands.

models of analysis—units and methods of studying public policy, offering specific perspectives on policymaking, implementation, and content.

natural disturbance regime—the natural frequency and intensity of disturbances, such as tornadoes, hurricanes, or complete wetland drying.

net primary production—biomass accumulated above and below the ground that accounts for loss through respiration (gross primary productivity − respiration = net primary productivity).

nitrification—the metabolic pathway where nitrifying bacteria convert ammonium to nitrate.

nitrifying bacteria—a specific class of bacteria in the genera *Nitrosomonas, Nitrosococcus,* and *Nitrobacter* that carry out nitrification.

nongame species—a species that is not of direct recreational or commercial value to humans; a species that is not harvested, hunted, or otherwise collected by humans for pleasure or economic gain.

non-point-source pollution—pollution resulting from many scattered locations without a single, defined source.

nontidal marsh—a subcategory of marsh, typically inland freshwater or brackish water marshes that are not affected by ocean tides and that may periodically dry up.

nonuse (passive-use) values—economic values enjoyed by people that do not require the active, on-site use of an area, wildlife specie, or artifact, often broken down into "existence," "bequest," and "option" values. For example, people may be willing to make some economic sacrifice in order to expand grizzly bear habitat and ensure that the grizzly bear survives as a species even though they have no desire to visit that grizzly bear habitat.

North American Wetlands Conservation Act—legislation designed in 1968 to achieve conservation of wetlands in Canada, Mexico, and the United States in connection with the protection of migratory birds.

northern bog—a specific type of bog found only in the Northeast and Great Lakes regions of the United States.

nutrient cycling—the movement of nutrients from elemental forms to biological forms through uptake and back to elemental forms through decomposition.

obligate anaerobes—a class of microorganisms that can use only anaerobic respiration or fermentation to produce energy because oxygen is toxic to them.

ombrotrophic—receiving minerals and nutrients exclusively from precipitation resulting in a relatively nutrient-poor wetland.

oxidation—a chemical process that involves the loss of electrons.

oxidation–reduction potential—the quantitative measure of electron availability in a chemical system.

partisan—characterized by support for a political party.

patch size—the surface area of a continuous area of habitat.

pH—the negative log of the hydrogen ion concentration; a quantitative measure of acidity.

physical environment—the nonliving component of ecosystems, including such factors as temperature, chemical composition of soils and water, and light availability.

physical takings—the traditional expropriation of private property, such as the confiscation of a specified amount of privately owned land in order to widen a street.

pluralism—a defining characteristic of a political system existing to manage and reflects competition between various interests, producing a balanced, consensual policy as a result of compromise between interest groups.

pocosin—a special type of bog located in the southeastern United States and dominated by evergreen-shrub vegetation.

polis—a city-state; refers to a political society motivated by public interest.

political culture—the aggregate attitudes of a community or state regarding the legitimacy, purpose, and success of government and citizenship.

population—a group of organisms of the same species occupying a common area and usually isolated from other groups of the same species.

precipitation-dominated wetland—a type of wetland that receives most of its water inputs from precipitation as opposed to groundwater, streams, or runoff.

preferred habitat—the available habitat where a species has the highest reproductive rate per individual.

presentism—the inappropriate use of present-day concerns or beliefs to understand or sometimes to condemn or praise past societies.

primary production—the converting of nutrients and light energy into tissue by plants (*see also* net primary production).

private property—property owned by individuals and recognized as such by the legal system.

private property rights groups—organizations devoted to the protection and expansion of private property rights and the sanctity of private property.

privatization—the provision of a public service (for example, trash collection), previously supplied by a political institution, by the private sector.

public goods—a nondivisible program or service responding to a public need and provided by government to all members of society.

public investment—the allocation of public resources to a project or goal.

public policy—the laws and regulations based on the decisions of government in regard to specific public issues.

public works projects—the construction of major infrastructure elements and facilities to serve the public.

redox-active compounds—elements and compounds with unfilled outer shells that will transfer electrons.

redoximorphic features—soil properties associated with wetness that result from the reduction and oxidation of iron, manganese, and carbon compounds in the soil.

redox potential—commonly used abbreviation for oxidation–reduction potential.

reduction—a chemical process that involves the gain of electrons.

reference wetlands—wetlands that are relatively pristine systems thought to have high ecological integrity that are used in comparison with wetlands that are being assessed or monitored.

regulatory rule making—the process used by federal agencies to establish the legal guidelines for implementation of legislative statutes.

regulatory takings—the reduction or elimination of economic value and uses of private property as a result of compliance with government regulations.

relative abundance—the relative distribution of individuals among species in a community.

remediation—a term used by some ecologists to refer to the restoration of highly degraded areas.

restoration—the attempt to return a degraded ecosystem to pristine or natural conditions.

rhizosphere—the area surrounding and influenced by the plant root.

riparian buffers—a linear piece of land adjacent to a stream or river where natural vegetation is allowed to remain, with the purpose of preventing pollutants from entering streams and rivers.

River and Harbors Act of 1899—a federal policy designed to ensure accessibility to U.S. rivers and harbors and to facilitate navigation.

saline—the term that is used to refer to something that is salty or that contains salt (such as seawater). Salt concentration is measured in terms of salinity.

secondary production—the conversion of plant tissue into animal tissue by grazers.

sediment exchange sites—chemically active areas on the surface of soil particles, usually negatively charged, where ions can be attached or displaced.

sediment–litter compartment—an arbitrarily defined storage unit in natural ecosystems comprised of the sediments and decomposing organic matter on the sediment surface.

seed bank—the collection of viable seeds buried in the soil.

sequent occupance—the principle that the significance of the physical environment changes as attitudes, objectives, and technical abilities of the people living in it change.

social capital—the extent to which a community or nation consists of members who participate in the social and political life of that community or nation.

species richness—the number of species found in a community or sample of organisms from a community.

stare decisis—Latin for "let the decision stand"; used to indicate the American judicial system's reverence for precedent.

succession—the natural development of ecosystems over time, including shifts in plant and animal species and changes in the physical environment.

successional stages—the stages through which an ecosystem passes as it develops over time (*see also* succession).

swamp—A type of wetland that is dominated by woody vegetation (either shrubs or trees) and characterized by hydric soils during the growing season and standing water at other times of the year.

"swampbuster" provisions—the provisions of the 1988 Farm Bill that offer disincentives to the development of wetlands. Among other things, swamp-buster provisions decreased agricultural subsidies for farmers who did develop wetlands.

takings—the removal of property from private hands to government ownership or public use (*see also* Fifth Amendment).

thermodynamics—the study of energy transfers resulting from physical and chemical reactions.

tidal marsh—a type of coastal marsh that is alternately flooded and dried as a result of ocean tides.

tidal wetlands—those wetlands affected by the daily actions of the ocean tidal cycles, resulting in alternating periods of exposure to air and inundation with water with the ebbing and flowing of the tides.

translocated—describes the movement of soluble materials from one area to another.

travel cost method—a method to determine the economic value of recreational visits to a natural area or other site by studying how the probability of a visit varies with the travel costs associated with visits to that site.

U.S. Army Corps of Engineers—a federal bureaucracy dedicated to civil and military engineering projects; slightly less than 2% of the Corps's personnel are military.

U.S. Court of Federal Appeals—the court authorized to hear constitutional claims associated with federal statutes, executive regulations, and implied or expressed contractual claims with the U.S. government.

valence state—the overall charge of an atom based on the number of electrons in its outer shell.

valency—a positive or negative integer used to represent the combining capacity of an atom or molecule as determined by the number of electrons that it will lose, add, or share in a chemical reaction.

water budget—the water budget consists of the inputs and output of water from a wetlands.

waterfowl—birds associated with aquatic habitats, especially those that swim, including ducks and geese.

watershed—an area of land that drains to the same waterway, such as a common stream, a lake, or an ocean.

wetland biogeochemistry—the interdisciplinary study of chemical reactions in wetlands, involving both biological and geochemical processes.

wetland construction—a project in which a wetland is created where none previously existed, with the goal of providing ecosystem functions typically performed by a natural wetland.

wetland delineation—the science of defining the boundaries of wetland systems, often for regulatory purposes.

wetland-dependent species—species of organisms that require wetlands for the completion of their life cycle.

wetland enhancement—a project in which one or more wetland functions are enhanced in an existing wetland.

wetland functions—the ecosystem processes occurring within the wetland regardless of their value to humans.

wetland preservation—a project in which a fully functional wetland is protected from possible future threats.

wetland restoration—a project in which a wetland that was previously degraded or altered by human activity is restored.

wetland structure—the physical attributes of a wetland, such as soil and vegetation.

Wise Use movement—a coalition of interest groups, industries, and recreational associations with a general opposition to environmental regulations and dedicated to fewer restrictions of land use and more local control over public lands.

Index